Invertible Fuzzy Topological Spaces

Anjaly Jose · Sunil C. Mathew

Invertible Fuzzy Topological Spaces

Anjaly Jose
Department of Mathematics
St. Joseph's College Devagiri
Calicut, India

Sunil C. Mathew
Department of Mathematics
Deva Matha College Kuravilangad
Kottayam, India

ISBN 978-981-19-3688-3   ISBN 978-981-19-3689-0   (eBook)
https://doi.org/10.1007/978-981-19-3689-0

Mathematics Subject Classification: 54A40, 03E72

This Springer imprint is published by the registered company Springer Nature Singapore Pte Ltd.
The registered company address is: 152 Beach Road, #21-01/04 Gateway East, Singapore 189721,
Singapore

Preface

The introduction of fuzzy sets in the world of mathematics was a paradigm shift, as it represented the uncertainties of real-life systems. Developed by combining the ordered structure and topological properties, fuzzy topology plays a pivotal role in nurturing local as well as global nature of classical topological properties. Investigated and experimented in a fuzzy sense, the fuzzy sets are undoubtedly a revolutionary move.

This book is a detailed study of invertible fuzzy topological spaces. Here, the main attention is paid to inverting pairs in a fuzzy topological space, local to global fuzzy topological properties and the associated spaces of invertible fuzzy topological spaces. Invertible L-topological spaces are also studied here. The book encompasses a diverse learning on invertibility in fuzzy topological spaces and discusses at length the basic concepts of fuzzy topology and L-topology. The chapters are presented in a very systematic way with plethora of examples that the readers can easily comprehend the concept presented in a readable manner with simple notations.

Chapters Description

The book consists of seven chapters. Chapter 1 illustrates the motivation behind the study of invertible fuzzy topological spaces. In this chapter, all the basic definitions and results for the subsequent reading are given. But, it is recommended to the reader to have a basic idea of fuzzy sets and its operations and also basic concepts of classic topology.

Chapter 2 classifies certain fuzzy topological spaces based on homeomorphisms introducing the concept of N-fuzzy topological spaces, strongly homogeneous fuzzy topological spaces, H-fuzzy topological spaces, complete H-fuzzy topological spaces and H-fuzzy topological spaces of degree n. Besides, the relationship between them, their connection with homogeneous and completely homogeneous fuzzy topological spaces have also been investigated. Finally, the sums, subspaces and simple extensions of these fuzzy topological spaces are explored.

Chapter 3 concentrates on the basic nature of invertible as well as completely invertible fuzzy topological spaces. The role of fuzzy points on the invertibility of a fuzzy topological space is examined. Apart from certain classes of invertible fuzzy topological spaces, conditions for the invertibility of some other types of fuzzy topological spaces are also obtained. In addition, situations under which a given fuzzy set is not an inverting fuzzy set are explored. It is observed that every fuzzy topological space can be embedded into an invertible fuzzy topological space. Further, characterizations and basic properties of completely invertible fuzzy topological spaces are derived. Again, the relationship between homogeneity and invertibility is established. This includes sufficient conditions for the invertibility of a homogeneous fuzzy topological space. The homogeneity of the inverting set as a subspace guarantees the homogeneity of the parent fuzzy topological space. In general, a completely invertible fuzzy topological space need not be homogeneous and a strongly homogeneous fuzzy topological space need not even be invertible. Finally, the orbits in invertible fuzzy topological spaces are studied and characterized. It is noted that in completely invertible fuzzy topological spaces, the orbits are always dense. On the other hand, if a completely invertible fuzzy topological space is not homogeneous then the orbits are neither open nor closed. Further, in a completely invertible fuzzy topological space, if any orbit as a subspace is non-trivial, then it is completely invertible. Consequently, every completely invertible fuzzy topological space is the disjoint union of homogeneous dense subspaces in which every non-trivial subspace is completely invertible. If a fuzzy topological space (X, F) is invertible, then there exists an inverting fuzzy set g and an inverting map θ of (X, F). This g and θ together are called an inverting pair of (X, F).

Chapter 4 closely examines the structure of these inverting pairs and observes that whenever θ is an inverting map so is θ^{-1}. Further, it is noted that if (g, θ) is an inverting pair then $\theta(g) = \theta^{-1}(g)$, whenever g and $\theta(g)$ are not quasi-coincident. Also, if (g, e) is an inverting pair, where e is the identity map, then the condition $\frac{1}{2} \leq g$ gives a clear picture about the structure of g. Based on the inverting maps, two types of invertible fuzzy topological spaces are introduced and their characterizing properties are derived. In the case of completely invertible fuzzy topological spaces, this classification actually produces some significant results. Further, the collection of all completely invertible finite fuzzy topological spaces is classified into two.

Chapter 5 investigates mainly the effect of invertibility on certain fuzzy topological properties. Exploring the relation between invertibility and separation axioms, it is proved that weakly quasi-separated, quasi-separated, fuzzy quasi T_0 and T_1 properties of certain subspaces can be transferred to the parent fuzzy topological space with the help of invertibility. The regular, normal, separated and fuzzy T_2 properties of a subspace determined by a crisp open subset accounts for the same property of a completely invertible fuzzy topological space. Further, the effect of invertibility on the axioms of countability is examined and certain sufficient conditions for an invertible fuzzy topological space to satisfy the first and second axioms of countability are obtained. If the inverting set as a subspace is separable, then the parent space is also separable. The role of invertibility on compactness and connectedness of a fuzzy topological space is also explored. It is proved that a type 2 completely invertible compact fuzzy topological space is strongly compact. Also, the compactness, α-compactness and α^*-compactness properties of the closure of an inverting crisp set as a subspace are carried over to the parent fuzzy topological space. Further, in a completely invertible strongly compact fuzzy topological space, every open fuzzy set contains a compact fuzzy subset. Interestingly, a completely invertible fuzzy topological space containing an open fuzzy connected subset has at most two components.

Chapter 6 studies sums, subspaces and simple extensions of different types of invertible fuzzy topological spaces and examines whether they remain in the same type or not. Even though complete invertibility is not additive, invertibility is an additive property. It is also proved that being type 1 and type 2 are not hereditary properties. While a type 1 invertible fuzzy topological space remains type 1 under simple extensions, type 2 looses that nature under simple extensions. But, type 2 nature of a completely invertible fuzzy topological space is retained with simple extensions. A related problem is to investigate the invertibility of the associated spaces and exploration in that direction produced some significant results. Invertibility of the quotient spaces of an invertible fuzzy topological space is also examined. A thorough investigation on the invertibility of the product space of a family of invertible fuzzy topological spaces has been carried out focusing on different types of invertible fuzzy topological spaces.

Chapter 7 extends the concept of invertibility to L-topological spaces and obtains certain properties of invertible L-topological spaces. It has been proved that stratification preserves invertibility of an L-topological space. Further, certain properties of inverting pairs are investigated. We also introduce completely invertible L-topological spaces and pinpoint some of their characteristics. Finally, we introduce two different types of invertible L-topological spaces and study their properties in relation to sums, subspaces and simple extensions. We also investigate the relationship between invertibility and countability axioms in L-topological spaces. In this direction, we prove that first countable, second countable and separable properties of certain subspaces are transferable to the parent L-topological space with the help of invertibility. We further investigate the effect of invertibility on the separation axioms in L-topological spaces and obtain certain local to global properties of invertible L-topologies.

Acknowledgement

Placing on record, we are deeply indebted to our colleagues, friends and family for the timely help and encouragement during the preparation of this book. We are also grateful to the reviewers and editors for their valuable suggestions and comments

which improved the presentation of this book. The first author wishes to thank RUSA
Phase II, St. Joseph's College, Devagiri, for their support.

Calicut, Kerala   Anjaly Jose
January 2022   Sunil C. Mathew

Contents

About the Authors

Anjaly Jose is Assistant Professor at the Department of Mathematics, St. Joseph's College Devagiri, Calicut, Kerala, India. She is a research guide under the University of Calicut, Kerala. Awarded the doctoral degree from Mahatma Gandhi University Kottayam in 2013, she is a resource person and has a number of research publications to her credit. Her areas of interest include topology and fuzzy topology.

Sunil C. Mathew is Principal of Deva Matha College Kuravilangad, Kerala, India. He has more than 26 years of teaching experience at St. Thomas College Palai, Kerala, India. Awarded the doctoral degree in 2003 from Mahatma Gandhi University Kottayam, Kerala, he is a research guide of the same university. A resource person with more than 40 international publications to his credit, his interests chiefly lie in fuzzy topology and graph labeling.

About the Authors

Chapter 1
Motivation and Preliminaries

As an introduction, this chapter explains the motivation behind the detailed study of invertible fuzzy topological spaces. Also discusses the basic definitions and results needed for the subsequent development of the study.

1.1 Introduction

Mathematics is a free creation of the human mind, as Cantor, the inventor of the modern theory of sets, says, *"the essence of Mathematics resides in its freedom, in the freedom to create"* [On infinite, linear point-manifolds (sets), 1883]. And history judges these creations by their enduring beauty and by the extent to which they illuminate other mathematical ideas or the physical universe. However as Einstein remarks *"as far as the laws of mathematics refer to reality, they are not certain; and as far as they are certain, they do not refer to reality"* [Geometry and Experience, 1921], the present era demands a new development in mathematics which will be capable of explaining the uncertainties in real-life systems.

In 1965, Zadeh[1] introduced fuzzy sets from the observation that *"more often than not, the classes of objects encountered in the real physical world do not have precisely defined criteria of membership"*. This provides us formalized tools for dealing with the imprecision intrinsic to many problems and the uncertainties and vagueness related to the real-life systems. The capability of fuzzy sets to express gradual transitions from membership to non-membership and vice-versa has immense utility.

The concept of a fuzzy topology was first put forward by Chang[2] in 1968. Since then various aspects of classical topology were investigated and carried out in fuzzy sense by several authors of this field. In [3], Lowen introduced an alternate definition of fuzzy topology and the concept of fuzzy compactness as a generalization of compactness in topology. Lowen continued the investigation on fuzzy compactness in [4]. Many authors [2, 5–8] have presented various kinds of compactness in fuzzy topological spaces.

In 1967, Gougen [9] introduced the concept of L-set where L is a complete distributive lattice. Consequently the theory of fuzzy sets was extended to L-sets. Later Goguen [10] himself proposed the natural generalization of Chang's definition by substituting L-sets for fuzzy sets. In [11], the concept of fuzzy point was introduced and the results concerning local countability, separability and local compactness were obtained. Separation axioms in fuzzy topological spaces are discussed in [6, 12–17]. Later Pu and Liu [18] redefined a fuzzy point in such a way that an ordinary point was a special case of it and they also introduced the concept of a Q-neighbourhood. The notions of quotient spaces and product spaces are extended to the fuzzy setting in [19]. In [20], Azad studied weaker forms of fuzzy continuity in fuzzy topological spaces. The notions of subspace and sum were extended to fuzzy topological spaces in [12]. In [21], it is shown that every fuzzy topological space is topologically isomorphic with a certain topological space. Several authors [22–25] also discussed different forms of connectedness in fuzzy topological spaces.

The group of homeomorphisms in a fuzzy topological space was studied by Johnson [26, 27]. Later, Mathew and Johnson [28] extended the concepts of g-closed sets and simple extensions to fuzzy topological spaces and established the relevance of these concepts in a fuzzy topological space. Strongly generalized closed fuzzy sets were introduced in [29] as a generalization of closed fuzzy set in fuzzy topological spaces. To be brief, the development of the non-trivial aspects of topology to the fuzzy setting is very impressive.

The concept of invertible topological spaces was introduced by Doyle and Hocking [30] in 1961, as a consequence of their characterization of the n-sphere. Later many have contributed to the development of invertible spaces. Gray [31] has proved that if an invertible space possesses a non-empty open subspace which is metrizable, then the space is metrizable. The effect of invertibility on separation axioms, investigated in [30], has been further explored by Levine [32] obtaining some local properties which are necessarily global in invertible spaces. Simultaneously, Wong undertook a detailed study on invertible spaces and improved certain results by weakening the conditions and obtained simpler proofs in [33]. Further, Umen [34] exhibited some properties of orbits in invertible spaces.

Doyle and Hocking [35] continuing their investigation on invertibility introduced the concept of continuously invertible topological spaces. Doyle and Hocking also introduced dimensional invertibility and obtained its applications on the theory of manifolds in [36]. Naimpally opened a quick glimpse on the function spaces of invertible spaces in [37]. The concept of generalized invertible spaces as introduced and studied by Hong in [38] and [39]. Ryeburn defined invertibility analogously for uniform spaces and showed that the whole space possesses certain properties holding for subspaces [40]. Hildebrand and Poe examined separation axioms for invertible spaces in [41]. The concept of semi-invertible spaces was introduced by Crossley and Hildebrand in [42]. Invertibility in infinite-dimensional spaces was thoroughly investigated by Tseng and Wong [43].

In 2006, Mathew [44] extended the concept of invertibility to fuzzy topological spaces and examined the basic nature of such spaces. In [45], Seenivasan and Balasubramanian discussed some properties of invertible fuzzy topological spaces.

In this book, a detailed study on invertible fuzzy topological spaces is conducted. The basic nature of such spaces is examined and as a result the role of fuzzy points in invertibility of a fuzzy topological space is obtained. The properties of completely invertible fuzzy topological spaces are investigated and characterized. Since invertibility depends on the group of homeomorphism of a space, the relation between homogeneity and invertibility in a fuzzy topological space is studied. Since orbits are homogeneous, the orbits in invertible and completely invertible fuzzy topological spaces are studied.

The study on the structure of inverting pairs in an invertible fuzzy topological space revealed the importance of the concept of quasi-coincidence in an invertible fuzzy topological space. If identity is an inverting map, a clear picture of the corresponding inverting fuzzy set will be obtained. This led to the classification of invertible fuzzy topological spaces into two types. In the case of completely invertible fuzzy topological spaces, this classification produces some significant results.

Having discussed the basic properties, the effect of invertibility on fuzzy topological properties is investigated and as a result the local to global nature of certain separation axioms, axioms of countability, compactness and connectedness in invertible and completely invertible fuzzy topological spaces are obtained. The crucial role of invertibility in fuzzy topological spaces is thus established.

Then it is only reasonable to investigate the related spaces of invertible fuzzy topological spaces. Subsequently, the sums, subspaces, simple extensions, associated spaces, quotient spaces and product spaces of different types of invertible fuzzy topological spaces are studied and examined whether invertibility is retained with these spaces or not.

Finally, the concept of invertibility is extended to L-topological spaces and examined the validity and relevance of the previous results on a bigger canvas.

1.2 Fuzzy Topological Spaces

Throughout this book X stands for a non-empty set with at least two elements and I stands for the unit interval $[0, 1]$. The identity map on X is denoted by e. The cardinality of a set A is denoted by $|A|$.

For any fuzzy subset g of X, by $\mathcal{C}(g)$ we mean the complement of g in X. A fuzzy set with constant degree of membership α is denoted by $\underline{\alpha}$. A fuzzy subset g of X is said to be proper if $g \neq \underline{0}, \underline{1}$.

Definition 1.1 [2] A fuzzy topology on a set X is a family F of fuzzy subsets of X, which satisfies the following three conditions:

(i) $\underline{0}, \underline{1} \in F$,
(ii) If $g, h \in F$ then $g \wedge h \in F$,

(iii) If $f_i \in F$ for each $i \in \Delta$ then $\bigvee_{i \in \Delta} f_i \in F$.

The pair (X, F) is called a fuzzy topological space or fts in short. The elements of F are said to be open fuzzy subsets of X and their complements are called closed fuzzy subsets of X. An fts (X, F) where $F = \{\underline{0}, \underline{1}\}$ is said to be trivial.

Definition 1.2 [21] Let f be a fuzzy subset of an fts (X, F). The closure of f is defined to be the fuzzy subset $\bigwedge \{g : g \geq f, \mathcal{C}(g) \in F\}$ of X and is denoted by $\overline{f}^X$ or simply $\overline{f}$.

Definition 1.3 [2] Let X and Y be two sets and let $\theta : X \to Y$ be a function. Then for any fuzzy subset g of X, $\theta(g)$ is a fuzzy subset in Y defined by

$$\theta(g)(y) = \begin{cases} \sup\{g(x) : x \in X, \ \theta(x) = y\}; \ \theta^{-1}(y) \neq \emptyset \\ 0; \qquad\qquad\qquad\qquad\qquad \theta^{-1}(y) = \emptyset \end{cases}$$

For a fuzzy subset h in Y, we define $\theta^{-1}(h)(x) = h(\theta(x))$, $\forall x \in X$. Obviously $\theta^{-1}(h)$ is a fuzzy subset of X.

Definition 1.4 [2] Let (X, F) and (Y, G) be two fuzzy topological spaces. Then a function $\theta : X \to Y$ is said to be continuous if $\theta^{-1}(g) \in F$ for every $g \in G$ and θ is said to be open if $\theta(f) \in G$ for every $f \in F$.

Definition 1.5 [2] Let (X, F) and (Y, G) be two fuzzy topological spaces. Then a bijection $\theta : X \to Y$ is said to be a homeomorphism if both θ and θ^{-1} are continuous. By a homeomorphism of (X, F) we mean a homeomorphism from (X, F) to itself.

Theorem 1.1 [46] *Let θ be a homeomorphism of (X, F). Then $\theta(\mathcal{C}(f)) = \mathcal{C}(\theta(f))$ for every fuzzy subset f of X.*

Definition 1.6 [18] The set $\{x \in X : f(x) > 0\}$ is called the support of f and is denoted by $supp\ f$.

Definition 1.7 [18] The fuzzy subset x_λ of X, with $x \in X$ and $0 < \lambda \leq 1$ defined by

$$x_\lambda(y) = \begin{cases} \lambda; & y = x \\ 0; & \text{otherwise} \end{cases}$$

is called a fuzzy point in X with support x and value λ. Two fuzzy points with different supports are called distinct.

Definition 1.8 [47] A fuzzy point x_λ is called weak or strong according as $\lambda \leq \frac{1}{2}$ and $\lambda > \frac{1}{2}$.

Definition 1.9 [47] Let (X, F) be an fts. A fuzzy point x_λ in X is said to be well-closed if $\overline{x_\lambda}$ is again a fuzzy point in X.

Definition 1.10 [47] Let (X, F) be an fts. A fuzzy subset g of X is said to be generalized closed (g-closed for brevity) if and only if $\overline{g} \leq f$ whenever $g \leq f$ and $f \in F$.

Theorem 1.2 [47] *Let* (X, F) *be an fts. Let* x_λ *and* x_γ *be two fuzzy points such that* $\lambda < \gamma$ *and* x_γ *is open. Then* x_λ *is well-closed if it is* g-*closed.*

Definition 1.11 [28] Let (X, F) be an fts. A fuzzy subset g of X is said to be nearly crisp if $\overline{g} \wedge \mathcal{C}(\overline{g}) = \underline{0}$.

Definition 1.12 [48] Let (X, F) be an fts. Then it is called a c-fts if the members of F are characteristic functions.

Definition 1.13 [49] An fts (X, F) is called a static fts if every $f \in F$ and $f \neq \underline{0}$ contains a nearly crisp fuzzy point.

Definition 1.14 [49] An fts (X, F) is called a D-fts if $f \wedge h \neq \underline{0}$ for every $f, h \in F$ and $f, h \neq \underline{0}$.

Definition 1.15 [49] Let (X, F) be a static fts. Let B be a smallest subset of X such that any $f \in F$ and $f \neq \underline{0}$ contains a nearly crisp fuzzy point with support in B. Such a set B is called a gem of the static fts (X, F) and $|B|$, the cardinality of B is defined to be the power of (X, F).

Definition 1.16 [19] An fts (X, F) is called fully stratified if $\underline{\alpha} \in F$, $\forall \alpha \in I$. An fts (X, F) is called purely stratified if whenever $f \in F$ then $f = \underline{\alpha}$ for some $\alpha \in I$.

Definition 1.17 [11] Let F be a fuzzy topology. A subfamily $\mathcal{B}$ of F is a base for F if each member of F can be expressed as the join of some members of $\mathcal{B}$.

Theorem 1.3 [11] *Let* (X, F) *be an fts. Then a subfamily* $\mathcal{B}$ *of F forms a base for F if and only if for every member* f *of F and for every fuzzy point* $x_\lambda \in f$, *there exists a member* h *of* $\mathcal{B}$ *such that* $x_\lambda \in h \leq f$.

Definition 1.18 [18] A fuzzy subset f in (X, F) is said to be a neighbourhood of x_λ if there exists an $h \in F$ such that $x_\lambda \in h \leq f$. The family consisting of all the neighbourhoods of x_λ is called the system of neighbourhoods of x_λ.

Definition 1.19 [18] Let $\mathcal{U}$ be a neighbourhood system of a fuzzy point x_λ in (X, F). A subfamily $\mathcal{B}$ of $\mathcal{U}$ is called a neighbourhood base of $\mathcal{U}$ if and only if for each $f \in \mathcal{U}$ there exists a member $g \in \mathcal{B}$ such that $g \leq f$.

Definition 1.20 [18] A fuzzy subset f is said to be quasi-coincident with the fuzzy subset g, denoted by $f \hat{q} g$ if and only if there exists $x \in X$ such that $f(x) + g(x) > 1$.

Theorem 1.4 [18] *Let* X *be any set and* f, g *be fuzzy subsets of* X. *Then* $f \leq g$ *if and only if* f *and* $\mathcal{C}(g)$ *are not quasi-coincident.*

Definition 1.21 [18] A fuzzy subset f in (X, F) is called a Q-neighbourhood of x_λ if there exists an $h \in F$ such that $x_\lambda \hat{q} h$ and $h \leq f$. The family consisting of all the Q-neighbourhoods of x_λ is called the system of Q-neighbourhoods of x_λ.

Definition 1.22 [18] Let $\mathcal{U}_Q$ be a Q-neighbourhood system of a fuzzy point x_λ in (X, F). A subfamily $\mathcal{B}_Q$ of $\mathcal{U}_Q$ is called a Q-neighbourhood base of $\mathcal{U}_Q$ if and only if for each $f \in \mathcal{U}_Q$ there exists a member $h \in \mathcal{B}_Q$ such that $h \leq f$.

Definition 1.23 [14] An fts (X, F) is said to be a T_1 fts if for any two distinct fuzzy points x_λ and y_γ there exists an $f \in F$ such that $x_\lambda \in f$ and $y_\gamma \notin f$ and another $g \in F$ such that $y_\gamma \in g$ and $x_\lambda \notin g$.

Theorem 1.5 [47] *Let (X, F) be an fts and x_λ be a g-closed but not well-closed fuzzy point. Then F is not T_1.*

Definition 1.24 [12] An fts (X, F) is said to be FT_1 if and only if for every pair of distinct fuzzy points x_λ and y_γ, there exist $f, g \in F$ such that $x_\lambda \in f \leq \mathcal{C}(y_\gamma)$ and $y_\gamma \in g \leq \mathcal{C}(x_\lambda)$.

Theorem 1.6 [28] *An fts (X, F) is FT_1 if and only if every fuzzy point in X is well-closed.*

Theorem 1.7 [6] *An fts is FT_1 if and only if every fuzzy point x_λ; $\lambda = 1$ is closed.*

Definition 1.25 [50] An fts (X, F) is said to satisfy $T_{1\frac{1}{2}}$-axiom if for any two fuzzy points x_λ and y_λ with $x_\lambda \in \mathcal{C}(y_\lambda)$, there exist f and g in F such that $f \leq \mathcal{C}(g)$, $x_\lambda \in f \leq \mathcal{C}(y_\lambda)$ and $y_\lambda \in g \leq \mathcal{C}(x_\lambda)$ or $y_\lambda \in f \leq \mathcal{C}(x_\lambda)$ and $x_\lambda \in g \leq \mathcal{C}(y_\lambda)$.

Definition 1.26 [17] An fts (X, F) is called quasi-separated if for any two fuzzy points x_λ and y_γ in X with $x_\lambda \leq \mathcal{C}(y_\gamma)$, there exist $f, g \in F$ such that $x_\lambda \in f \leq \mathcal{C}(y_\gamma)$ and $y_\gamma \in g \leq \mathcal{C}(x_\lambda)$.

Theorem 1.8 [17] *An fts (X, F) is quasi-separated if and only if every fuzzy point in X is closed.*

Definition 1.27 [12] An fts (X, F) is called regular if for any fuzzy point x_λ in X and any closed fuzzy subset h of X such that $x_\lambda \in \mathcal{C}(h)$, there exist $f, g \in F$ such that $x_\lambda \in f$ and $h \leq g$ with $f \leq \mathcal{C}(g)$.

Theorem 1.9 [28] *Let (X, F) be an fts in which every fuzzy subset is g-closed. Then (X, F) is regular.*

Definition 1.28 [50] A fuzzy topological spaces (X, F) is said to satisfy strongly regular axiom if given a fuzzy point x_λ and a closed set h in (X, F) with $x_\lambda \in \mathcal{C}(h)$, there exist $f \in F$ such that $x_\lambda \notin f$ and $h \leq f$.

Definition 1.29 [12] An fts (X, F) is called normal if for every two closed fuzzy subsets h_1 and h_2 such that $h_1 \leq \mathcal{C}(h_2)$, there exist two open fuzzy subsets f_1 and f_2 such that $h_1 \leq f_1$ and $h_2 \leq f_2$ and $f_1 \leq \mathcal{C}(f_2)$.

Definition 1.30 [50] A fuzzy topological space (X, F) is said to be completely normal if every subspace of (X, F) is normal.

Definition 1.31 [28] A fuzzy subset g of X is said to be strong if for every $x \in X$, g contains a strong fuzzy point x_λ. A fuzzy subset g of X is said to be weak if $\mathcal{C}(g)$ is a strong fuzzy subset of X not containing a crisp singleton.

Definition 1.32 [44] Let g be a fuzzy subset of X. Then the strong support of g is defined as $ssg = \{x \in X : g(x) > \frac{1}{2}\}$ and weak support of g is defined as $wsg = \{x \in X : g(x) \leq \frac{1}{2}\}$.

Definition 1.33 [51] Let (X, F) be an fts. Then a fuzzy subset f of X is said to be dense if $\overline{f} = 1$.

Definition 1.34 [3] For a topology τ on X let $\omega(\tau)$ be the subset of all lower semi-continuous functions from (X, τ) to $[0, 1]$. Then $\omega(\tau)$ turns out to be a fuzzy topology on X called the associated fuzzy topology of (X, τ). A fuzzy topology of the form $\omega(\tau)$ is said to be topologically generated. For a fuzzy topology F on X, $i(F)$ is the topology on X induced by all functions $f : X \to I_r$, where $f \in F$ and $I_r = [0, 1]$ with subspace topology of right ray topology on R.

Theorem 1.10 [26] *Let (X, F) be any fts. Then the set of all homeomorphisms $G(F)$ of (X, F) forms a group.*

Theorem 1.11 [26] *Let (X, F) be any fts. Then the group of homeomorphisms of (X, F) is a subgroup of the group of homeomorphisms of $(X, i(F))$.*

Theorem 1.12 [26] *If (X, F) is topologically generated, then the group of homeomorphisms of (X, F) and the group of homeomorphisms of $(X, i(F))$ are the same.*

Definition 1.35 [11] An fts (X, F) is said to be separable if there exists a countable sequence of fuzzy points $\{p_i\}$, $i = 1, 2, ...$, such that for every $f \in F$ and $f \neq 0$, there exists a p_i such that $p_i \in f$.

Definition 1.36 [11] Let $p = x_\lambda$ be a fuzzy point in (X, F). Let f be a fuzzy set in X. Then x_λ is an accumulation point of f if for every member g of F such that $x_\lambda \in g$, $g \wedge f_p \neq 0$, where f_p is a fuzzy subset of X defined by

$$f_p(y) = \begin{cases} 0; & y = x \\ f(y); & \text{otherwise} \end{cases}$$

Definition 1.37 [12] Let (X_i, F_i), $i \in \Delta$ be a family of pair wise disjoint fuzzy topological spaces. Define $X = \bigcup_{i \in \Delta} X_i$ and $F = \{f \in I^X : f \wedge X_i \in F_i, \text{ for all } i \in \Delta\}$. Then F is a fuzzy topology on X called the sum fuzzy topology on X and (X, F) is called the sum fts of the family (X_i, F_i), $i \in \Delta$.

Definition 1.38 [12] A fuzzy topological property ρ is additive if the sum fts (X, F) of every family (X_i, F_i), $i \in \Delta$ of pairwise disjoint fts enjoying property ρ also has property ρ.

Definition 1.39 [12] Let (X, F) be an fts and let M be any crisp subset of X. Then the induced fuzzy topology or relative fuzzy topology for M is given by $F_M = \{M \wedge f : f \in F\}$ and the pair (M, F_M) is called a subspace of (X, F).

Definition 1.40 [12] A fuzzy topological property ρ is called hereditary, if each subspace of an fts with property ρ also has property ρ.

Definition 1.41 [28] Let (X, F) be an fts and suppose that $g \in I^X$ and $g \notin F$. Then the collection $F(g) = \{g_1 \vee (g_2 \wedge g) : g_1, g_2 \in F\}$ is called the simple extension of F determined by g.

Definition 1.42 [23] If f and g are fuzzy subsets of X, then $f + g$ is the function, $f + g : X \to [0, 2]$ such that $(f + g)(x) = f(x) + g(x)$ for all $x \in X$.

Note 1.1 Let (X, F) be an fts. Let $\Lambda_F = \{\lambda \in [0, 1] : f(x) = \lambda$, for some $f \in F$ and $x \in X\}$. For each $\lambda \in [0, 1]$, let $X_\lambda = \{x \in X : f(x) = \lambda$ for some $f \in F$ such that $f \neq \underline{0}, \underline{1}\}$ and $F_\lambda = \{f \in F : f(x) = \lambda$ for some $x \in X\}$. Also for each $g \in F$ and $\alpha \in I$, let $S_\alpha(g) = \{x \in X : g(x) = \alpha\}$.

1.3 L-Topological Spaces

L-subset with constant degree of membership α is denoted by $\underline{\alpha}$. The power set of a set X is denoted by $P(X)$. An L-subset v of X is said to be proper if $v \neq \underline{0}, \underline{1}$. The least infinite ordinal number is denoted by ω.

Definition 1.43 [52] A completely distributive lattice L is called an F-lattice, if L has an order-reversing involution $' : L \to L$. Let X be a non-empty ordinary set, L an F-Lattice, $'$ the order-reversing involution on L. For any $u \in L^X$, use the order-reversing involution $'$ to define an operation $'$ on L^X by: $u'(x) = (u(x))'$ $\forall x \in X$. We call $' : L^X \to L^X$ the pseudo-complementary operation on L^X, u' the pseudo-complementary set of $u \in L^X$.

Theorem 1.13 [52] *Let X be a non-empty ordinary set, L an F-lattice, then the pseudo-complementary operation $' : L^X \to L^X$ is an order-reversing involution.*

Definition 1.44 [52] Let X be a non-empty ordinary set, L an F-lattice and $\delta \subset L^X$. Then δ is called an L-topology on X, and (L^X, δ) is called an L-topological space, or L-ts for short, if δ satisfies the following three conditions:

(i) $\underline{0}, \underline{1} \in \delta$,
(ii) $\forall u, v \in \delta, u \wedge v \in \delta$,
(iii) $\forall \mathcal{A} \subset \delta, \vee \mathcal{A} \in \delta$.

Every element in δ is called an open L-subset in L^X and every pseudo-complementary set of an open L-subset is called a closed L-subset in L^X. The smallest closed L-subset containing u is called the closure of u and is denoted by u^-. An L-ts (L^X, δ) where $\delta = \{\underline{0}, \underline{1}\}$ is said to be trivial.

Definition 1.45 [52] Let L^X, L^Y be two L-spaces, $\theta : X \to Y$ an ordinary mapping. Based on $\theta : X \to Y$, define L-mapping $\theta^\to : L^X \to L^Y$ by $\theta^\to(v)(y) = \sup\{v(x) : x \in X, \theta(x) = y\}$; $\forall v \in L^X$, $\forall y \in Y$ and its L-reverse mapping $\theta^\gets : L^Y \to L^X$ by $\theta^\gets(w)(x) = w(\theta(x))$, $\forall w \in L^Y$, $\forall x \in X$.

Definition 1.46 [52] Let (L^X, δ), (L^Y, μ) be L-ts, $\theta^\to : L^X \to L^Y$ an L-mapping. (i) $\theta^\to : (L^X, \delta) \to (L^Y, \mu)$ is called an L-continuous mapping from (L^X, δ) to (L^Y, μ) if $\forall v \in \mu, \theta^\gets(v) \in \delta$. (ii) $\theta^\to : (L^X, \delta) \to (L^Y, \mu)$ is called open, if $\forall u \in \delta$, $\theta^\to(u) \in \mu$. (iii) $\theta^\to : (L^X, \delta) \to (L^Y, \mu)$ is called an L-homeomorphism, if it is bijective continuous and open.

Theorem 1.14 [53] *Let (L^X, δ), (L^Y, μ) be L-ts, $\theta^\to : L^X \to L^Y$ an L-homeomorphism, then for every $u \in L^X$, $\theta^\to(u') = (\theta^\to(u))'$.*

Theorem 1.15 [52] *Let (L^X, δ), (L^Y, μ) be L-ts, $\theta^\gets : L^Y \to L^X$ an L-homeomorphism, then for every $v \in L^Y$, $\theta^\gets(v') = (\theta^\gets(v))'$.*

Definition 1.47 [54] An L-point on X with support x and value $\lambda \neq 0$, is an L-subset $x_\lambda \in L^X$ defined as, for any $y \in X$,

$$x_\lambda(y) = \begin{cases} \lambda; & y = x \\ 0; & \text{otherwise} \end{cases}$$

The set of all L-points on X is denoted by $Pt(L^X)$. For every $\mathcal{A} \subset L^X$, denote the set of all L-points on X in $\mathcal{A}$ by $Pt(\mathcal{A})$.

Definition 1.48 [52] A join-irreducible element of L is called a molecule in L. For every $A \subset L$, the set of all molecules of L in A is denoted by $M(A)$. The set of all the molecules of L^X in a family $\mathcal{A} \subset L^X$, denoted by $M(\mathcal{A})$, is defined by $M(\mathcal{A}) = \{x_\lambda \in Pt(A) : \lambda \in M(L)\}$.

Definition 1.49 [52] Let L^X be an L-space and $k \in L^X$, $a \in L$. Define the a-level of k as the ordinary set $\{x \in X : k(x) \geq a\}$, denoted by $k_{[a]}$. Also denote $k_{(a)} = \{x \in X : k(x) \nleq a\}$.

Definition 1.50 [52] Let L^X be an L-space, $a \in L - \{0\}$. $k \in L^X$ is called a-crisp, if $k_{(0)} = k_{[a]}$; k is called pseudo-crisp, if there exists $a \in L - \{0\}$ such that k is a-crisp.

Definition 1.51 [55] An L-point x_λ in X is called weak or strong according as $x_\lambda \in x_\lambda'$ and $x_\lambda \notin x_\lambda'$.

Definition 1.52 [55] Let (L^X, δ) be an L-ts. An L-point x_λ in X is said to be well-closed if x_λ^- is again an L-point in X.

Definition 1.53 [52] Let (L^X, δ) be an L-ts. $\forall x_\alpha \in L^X$, $\forall u, v \in L^X$, we say x_α quasi-coincident with u, denoted by $x_\alpha \hat{q} u$, if $x_\alpha \not\leq u'$ or $\alpha \not\leq u'(x)$ and say u quasi-coincident with v at x, if $u(x) \not\leq v'(x)$; say u is quasi-coincident with v if u quasi-coincident with v at some point $x \in X$.

Definition 1.54 [52] Let (L^X, δ) be an L-ts. (L^X, δ) is called typically T_1 if every L-point in X is a closed L-subset.

Definition 1.55 [52] For every L-subset $u \in L^X$, its support set denoted by *supp u* is defined as $\{x \in X : u(x) > 0\}$.

Definition 1.56 [52] Let (L^X, δ) be an L-ts. Then the L-topology μ on X generated by $\delta \cup \{\underline{a} : a \in L\}$ is called the stratification of δ and (L^X, μ) is called stratification of (L^X, δ).

Definition 1.57 [52] Let $S = \{(X_t, \delta_t), t \in T\}$ be a family of L-ts, $\mathcal{A} = \{u_t : t \in T\}$ a family of L-subsets where $u_t \in X_t$ for every $t \in T$. Denote $X = \prod_{t \in T} X_t$. For every $t \in T$, suppose $p_t : X \to X_t$ is the ordinary projection, define the projection from L-space L^X to L^{X_t} as $p_t : L^X \to L^{X_t}$. Define the product topology of L-topologies $\{\delta_t : t \in T\}$ on X, denoted by $\prod_{t \in T} \delta_t$, as the L-topology δ on X generated by the subbase $\{p_t^{-1}(u_t) : u_t \in \delta_t, t \in T\}$, and call the L-ts (X, δ) the product space of L-ts $\{(X_t, \delta_t), t \in T\}$. For every $t \in T$, (X_t, δ_t) is called a coordinate space. Define the product of L-subsets $\mathcal{A} = \{u_t : t \in T\}$, denoted by $\prod \mathcal{A} = \prod_{t \in T} u_t = \bigwedge \{p_t^{-1}(u_t) : t \in T\}$.

Theorem 1.16 [52] *Let* $\{(X_t, \delta_t) : t \in T\}$ *be a family of L-ts, (X, δ) be their product space. Then* $\{\bigwedge_{t \in F} p_t^{-1}(u_t) : F \in [T]^{<\omega}\}$ *is a base of the product topology δ.*

Note 1.2 Let (L^X, δ) be an L-ts. Then for each $\lambda \in L$,
let $X_\lambda = \{x \in X : u(x) = \lambda \text{ for some } u \in \delta \text{ such that } u \neq \underline{0}, \underline{1}\}$. Also for $v \in \delta$ and $\alpha \in L$, we define $S_\alpha(v) = \{x \in X : v(x) = \alpha\}$.

References

1. Zadeh, L.A.: Fuzzy sets. Inf. Control **8**, 338–355 (1965)
2. Chang, C.L.: Fuzzy topological spaces. J. Math. Anal. Appl. **24**, 182–190 (1968)
3. Lowen, R.: Fuzzy topological spaces and fuzzy compactness. J. Math. Anal. Appl. **56**, 621–633 (1976)
4. Lowen, R.: A comparison of different compatness notions in fuzzy topological spaces. J. Math. Anal. Appl. **64**, 446–454 (1978)
5. Gantner, R.C., Steinlage, T.E., Warren, R.H.: Compactness in fuzzy topological spaces. J. Math. Anal. Appl. **62**, 547–562 (1978)
6. Tsiporkova, E., Kerre, E.: On separation axioms and compactness in fuzzy topological spaces. J. Egypt. Math. Soc. **4**, 27–39 (1996)
7. Liu, Y.M.: The compactness and tychnoff product theorem in fuzzy topological spaces. Acta Math. Sinica **24**, 262–268 (1981)

8. Wang, G.J.: A new fuzzy compactness defined by fuzzy nets. J. Math. Anal. Appl. **94**, 1–23 (1983)
9. Goguen, J.A.: L-fuzzy sets. J. Math. Anal. Appl. **18**, 145–174 (1967)
10. Goguen, J.A.: The fuzzy tychnoff theorem. J. Math. Anal. Appl. **43**, 734–742 (1973)
11. Wong, C.K.: Fuzzy points and local properties of fuzzy topology. J. Math. Anal. Appl. **46**, 316–328 (1974)
12. Ghanim, M.H., Kerre, E.E., Mashhour, A.S.: Separation axioms, subspaces and sums in fuzzy topology. J. Math. Anal. Appl. **102**, 189–202 (1984)
13. Srivastva, S.N., Lal, R., Srivastava, A.K.: Fuzzy hausdorff topological spaces. J. Math. Anal. Appl. **81**, 497–507 (1981)
14. Srivastva, S.N., Lal, R., Srivastava, A.K.: Fuzzy T_1-topological spaces. J. Math. Anal. Appl. **127**, 1–7 (1987)
15. Hutton, B.: Normality in fuzzy topological spaces. J. Math. Anal. Appl. **50**, 74–79 (1975)
16. Ganguly, S., Saha, S.: On separation axiom and T_1 fuzzy continuity. Fuzzy Sets Syst. **50**, 265–275 (1985)
17. Tsiporkova, E., Baets, B.D.: τ-compactness in separated fuzzy topological spaces. Tatra Mt. Math. Publ. **12**, 99–112 (1997)
18. Pu, P.M., Liu, Y.M.: Fuzzy topology. 1. neighbourhood structure of a fuzzy point and moore-smith convergence. J. Math. Anal. Appl. **76**, 571–599 (1980)
19. Pu, P.M., Liu, Y.M.: Fuzzy topology. 11. product and quotient spaces. J. Math. Anal. Appl. **77**, 20–37 (1980)
20. Azad, K.K.: On fuzzy semicontinuity, fuzzy almost continuity and fuzzy weakly continuity. J. Math. Anal. Appl. **82**, 14–32 (1981)
21. Hu, C.M.: Fuzzy topological spaces. J. Math. Anal. Appl. **110**, 141–175 (1985)
22. Zheng, C.Y.: Fuzzy path and fuzzy connectedness. Fuzzy Sets Syst. **14**, 273–280 (1984)
23. Fatteh, U.V., Bassan, D.S.: Fuzzy connectedness and its stronger forms. J. Math. Anal. Appl. **III**, 449–464 (1985)
24. Lowen, R.: Connectedness in fuzzy topological spaces. Rocky Mt. J. Math. **11**(3), 427–433 (1981)
25. Wuyt, P.: Fuzzy path and fuzzy connectedness. Fuzzy Sets Syst. **24**, 127–128 (1987)
26. Johnson, T.P.: On the group of fuzzy homeomorphisms. J. Math. Anal. Appl. **175**, 333–336 (1993)
27. Johnson, T.P.: The group of fuzzy homeomorphisms. Fuzzy Sets Syst. **45**, 355–358 (1992)
28. Mathew, S.C., Johnson, T.P.: Generalized closed fuzzy sets and simple extensions of a fuzzy topology. J. Fuzzy Math. **11**(1), 195–202 (2003)
29. Zakari, A.H., Ai-Saadi, H.S.: Strongly generalazied closed fuzzy sets in fuzzy topological spaces and new separation axioms. J. Fuzzy. Math. **18**(1), 217–231 (2010)
30. Doyle, P.H., Hocking, J.G.: Invertible spaces. Am. Math. Mon. **68**(10), 959–965 (1961)
31. Gray, W.J.: On the metrizability of invertible spaces. Am. Math. Mon. **71**(5), 533–534 (1964)
32. Levine, N.: Some remarks on invertible spaces. Am. Math. Mon. **70**(2), 181–183 (1963)
33. Wong, Y.M.: Some remarks on invertible spaces. Am. Math. Mon. **73**(8), 835–841 (1966)
34. Umen, A.J.: Some remarks on orbits in invertible spaces. Am. Math. Mon. **71**(6), 643–646 (1964)
35. Doyle, P.H., Hocking, J.G.: Continuously invertible spaces. Pacific J. Math. **12**(2), 499–503 (1962)
36. Doyle, P.H., Hocking, J.G.: Dimensional invertibility. Pacific J. Math. **12**(2), 1236–1240 (1962)
37. Naimpally, S.A.: Function spaces of invertible spaces. Am. Math. Mon. **73**(5), 513–515 (1966)
38. Hong, D.X.: Generalized invertible spaces. Am. Math. Mon. **73**(2), 150–154 (1966)
39. Hong, D.X.: Some theorems on generalized invertible spaces. Am. Math. Mon. **75**(4), 378–382 (1968)
40. Ryeburn, D.: Finite additivity, invertibility, and simple extensions of topologies. Am. Math. Mon. **74**(2), 148–152 (1967)

41. Hildebrand, S.K., Poe, R.L.: The separation axioms for invertible spaces. Am. Math. Mon. **75**(4), 391–392 (1968)
42. Crossley, S.C., Hildebrand, S.K.: Semi-invertible spaces. Texas J. Sci. **26**, 319–324 (1975)
43. Tseng, C.C., Wong, N.C.: Invertibility in infinite-dimensional spaces. Proc. Am. Math. Soc. **128**(2), 573–581 (2000)
44. Mathew, S.C.: Invertible fuzzy topological spaces. STARS: Int. J. **7**, 1–13 (2006)
45. Seenivasan, V., Balasubramanian, G.: Invertible fuzzy topological spaces. Ital. J. Pure Appl. Math. **22**, 223–230 (2007)
46. Mathew, S.C., Jose, A.: On the structure of invertible fuzzy topological spaces. Advances Fuzzy Sets Syst. **40**(1), 67–79 (2010)
47. Johnson, T.P., Mathew, S.C.: On fuzzy points in fuzzy topological spaces. Far East J. Math. Sci. Sp. **1**, 75–86 (2000)
48. Mathew, S.C., Johnson, T.P.: On c-fuzzy topological space and simple extensions. J. Tri. Math. Soc. **3**, 55–58 (2001)
49. Mathew, S.C., Johnson, T.P.: Static fuzzy topological spaces. J. Fuzzy Math. **10**(4), 831–838 (2002)
50. Rao, K.C.: On invertible fuzzy topological spaces. J. Adv. Stud. Topol. **2**(1), 18–20 (2011)
51. Seenivasan, V., Balasubramanian, G.: Fuzzy semi α-irresolute functions. Math. Bohem. **132**(2), 113–123 (2007)
52. Liu, Y.M., Luo, M.K.: Fuzzy Topology. World Scientific (1997)
53. Jose, A., Mathew, S.C.: Invertibility in L-topological spaces. Fuzzy Inf. Eng. **6**, 41–57 (2014)
54. Jose, A., Mathew, S.C.: Local to global properties of invertible l topologies. Anals. U. Oradea-Fasicola Mathematica **2**, 85–94 (2013)
55. Mathew, S.C.: Adjacent L-fuzzy topological spaces. Int. J. Appl. Math. Stat. **7**, 1–13 (2006)

Chapter 2
H-Fuzzy Topological Spaces

This chapter classifies certain fuzzy topological spaces based on homeomorphisms introducing the concept of N-fuzzy topological spaces, strongly homogeneous fuzzy topological spaces, H-fuzzy topological spaces, complete H-fuzzy topological spaces and H-fuzzy topological spaces of degree n. Besides the relationship between them, their connection with homogeneous and completely homogeneous fuzzy topological spaces has also been investigated. Finally, the sums, subspaces and simple extensions of these fuzzy topological spaces are explored.

2.1 H-Fuzzy Topological Spaces

This section introduces different types of fuzzy topological spaces based on homeomorphism and studies the basic properties of H-fuzzy topological spaces. The sums, subspaces and simple extensions of it are also investigated.

Definition 2.1 [1] An fts (X, F) is called an N-fts if there exists a homeomorphism other than the identity on X.

Definition 2.2 [2] An fts (X, F) is said to be homogeneous if for every pair of elements $x, y \in X$ there exists a homeomorphism θ of (X, F) such that $\theta(x) = y$.

Definition 2.3 [3] An fts (X, F) is said to be strongly homogeneous if for every pair of elements $x, y \in X$ there exists a homeomorphism θ such that $\theta(x) = y$ and $\theta(y) = x$.

Definition 2.4 [4] An fts (X, F) is said to be completely homogeneous if every bijection on X is a homeomorphism of (X, F).

Definition 2.5 [5] An fts (X, F) is said to be an H-fts if for any $x \in X$, there exists a homeomorphism θ of (X, F) such that $\theta(x) \neq x$. An H-fts is said to be complete if there exists a homeomorphism θ of (X, F) such that $\theta(x) \neq x$, $\forall x \in X$.

© The Author(s), under exclusive license to Springer Nature Singapore Pte Ltd. 2022   13
A. Jose and S. C. Mathew, *Invertible Fuzzy Topological Spaces*,
https://doi.org/10.1007/978-981-19-3689-0_2

Example 2.1 Let X be any set. Choose $\lambda \in I$. Define $G = \{f \in I^X : f(x) = \lambda \text{ for}$ some $x \in X\}$. Let F be the fuzzy topology generated by G. Then (X, F) is an H-fts.

Example 2.2 Let X be the set of all non-zero real numbers. Define $G = \{f \in I^X : \text{if}$ $f(x) = \alpha > 0$ then there exists an $\varepsilon > 0$ such that $f(y) = \alpha$, $\forall y \in (x - \varepsilon, x + \varepsilon)\}$. Let F be the fuzzy topology generated by G. Define $\theta : X \to X$ defined by $\theta(x) = -x$. Clearly θ is a homeomorphism on X such that $\theta(x) \neq x$, $\forall x \in X$. Consequently (X, F) is a complete H-fts.

The following theorem sheds some light on the structure of open fuzzy subsets in an H-fuzzy topological space.

Theorem 2.1 [5] *For an H-fts* (X, F), $|X_\alpha| \geq 2$, $\forall \alpha \in \Lambda_F$.

Proof Let $\alpha \in \Lambda_F$. Then there exists an $x \in X$ and $f \in F$ such that $f(x) = \alpha$. Since (X, F) is an H-fts, there exists a homeomorphism θ such that $\theta(x) = x_0$ for some $x_0 \in X - \{x\}$. But $\theta(f)(x_0) = f(x) = \alpha$. Consequently $|X_\alpha| \geq 2$. $\square$

Remark 2.1 Obviously every H-fts is an N-fts. But the converse is not true as verified by the following example.

Example 2.3 Let $X = \{p, q, r, s\}$. Consider $f, g \in I^X$ defined by

$$f(p) = \tfrac{2}{3}, \ f(q) = \tfrac{1}{3}, \ f(r) = 0, \ f(s) = 0$$

$$g(p) = \tfrac{1}{3}, \ g(q) = \tfrac{1}{3}, \ g(r) = 0, \ g(s) = 0.$$

Then (X, F) is an fts where $F = \{\underline{0}, \underline{1}, f, g\}$. Consider $\theta : X \to X$ defined by $\theta(p) = p$, $\theta(q) = q$, $\theta(r) = s$, $\theta(s) = r$. Clearly θ is a homeomorphism of (X, F) other than the identity. Consequently (X, F) is an N-fts. But there exists no homeomorphism η of (X, F) such that $\eta(p) \neq p$. Thus (X, F) is not an H-fts.

Theorem 2.2 [5] *A homogeneous fts is always an H-fts.*

Proof Suppose (X, F) be homogeneous and let $x \in X$. Then for any $y \in X$ with $y \neq x$ there exists a homeomorphism θ such that $\theta(x) = y$. Thus for any $x \in X$, there exists a homeomorphism θ such that $\theta(x) \neq x$. Therefore (X, F) is an H-fts. $\square$

Remark 2.2 Converse of the above theorem is not true. For example, let $X = \{a, b, c, d\}$ and consider the fuzzy subsets f, g, h of X defined by

$$f(a) = \tfrac{1}{3}, \quad f(b) = \tfrac{1}{3}, \quad f(c) = 0, \quad f(d) = 0$$

$$g(a) = 0, \quad g(b) = 0, \quad g(c) = \tfrac{1}{2}, \quad g(d) = \tfrac{1}{2}$$

$$h(a) = \tfrac{1}{3}, \quad h(b) = \tfrac{1}{3}, \quad h(c) = \tfrac{1}{2}, \quad h(d) = \tfrac{1}{2}$$

Let $F = \{\underline{0}, \underline{1}, f, g, h\}$. Then (X, F) is an fts. Define $\theta : X \to X$ by $\theta(a) = b$, $\theta(b) = a$, $\theta(c) = d$, $\theta(d) = c$. Then it follows that θ is a homeomorphism and

$\theta(x) \neq x$, $\forall x \in X$. Therefore (X, F) is a complete H-fts. But (X, F) is not homogeneous because there exists no homeomorphism of (X, F) which maps $b \to c$. Thus there are complete H-fuzzy topological spaces which are not homogeneous. But these two notions coincide when $|X| = 2$.

Remark 2.3 The bijective continuous image of an H-fts need not be an H-fts. For example, let $X = \{a, b, c, d\}$ and $Y = \{p, q, r, s\}$. Consider the fuzzy sets f, g, h of X and i of Y where

$$f(a) = \tfrac{1}{6}, \quad f(b) = \tfrac{5}{6}, \quad f(c) = 0, \quad f(d) = 0$$

$$g(a) = 0, \quad g(b) = 0, \quad g(c) = \tfrac{1}{6}, \quad g(d) = \tfrac{5}{6}$$

$$h(a) = \tfrac{1}{6}, \quad h(b) = \tfrac{5}{6}, \quad h(c) = \tfrac{1}{6}, \quad h(d) = \tfrac{5}{6}$$

$$i(p) = \tfrac{1}{6}, \quad i(q) = \tfrac{5}{6}, \quad i(r) = 0, \quad i(s) = 0$$

Let $F = \{\underline{0}, \underline{1}, f, g, h\}$. Then (X, F) is an H-fts. Now (Y, G) is an fts where $G = \{\underline{0}, \underline{1}, i\}$. Define $\eta : X \to Y$ as $\eta(a) = p$, $\eta(b) = q$, $\eta(c) = r$, $\eta(d) = s$. Then $\eta^{-1}(i) = f \in F$ so that η is continuous. But it can be easily verified that (Y, G) is not an H-fts.

The following example shows that the continuous, open and surjective image of an H-fts need not be an H-fts.

Example 2.4 Let $X = \{a, b, c, d\}$. Consider $f, g, h \in I^X$ defined by

$$f(a) = \tfrac{1}{8}, \quad f(b) = \tfrac{7}{8}, \quad f(c) = \tfrac{1}{8}, \quad f(d) = \tfrac{7}{8}$$

$$g(a) = 0, \quad g(b) = 0, \quad g(c) = \tfrac{1}{8}, \quad g(d) = \tfrac{7}{8}$$

$$h(a) = \tfrac{1}{8}, \quad h(b) = \tfrac{7}{8}, \quad h(c) = 0, \quad h(d) = 0.$$

Then (X, F) is an H-fts where $F = \{\underline{0}, \underline{1}, f, g, h\}$. Let $Y = \{p, q\}$ and consider the fuzzy subset i of Y defined by $i(p) = \tfrac{1}{8}$, $i(q) = \tfrac{7}{8}$. Then (Y, G) is an fts where $G = \{\underline{0}, \underline{1}, i\}$. Now define $\theta : X \to Y$ by $\theta(a) = p$, $\theta(b) = q$, $\theta(c) = p$, $\theta(d) = q$. Then $\theta^{-1}(i) = f \in F$ so that θ is continuous. Also it follows that $\theta(f) = \theta(g) = \theta(h) = i$. Thus θ is continuous, open and surjective. But (Y, G) is not an H-fts.

Theorem 2.3 [5] *If* (X, F) *is an H-fts, then* $(X, i(F))$ *is also an H-fts.*

Proof Let $x \in X$. Since (X, F) is an H-fts there exists a homeomorphism θ of (X, F) such that $\theta(x) \neq x$. Then by theorem 1.12, θ is a homeomorphism of $(X, i(F))$ also. Hence $(X, i(F))$ is an H-fts. $\qquad\square$

Remark 2.4 Converse of the above theorem is not true. For example, let $X = \{a, b, c, d\}$. Consider the fuzzy subsets f, g, h of X defined by

$$
\begin{array}{llll}
f(a) = \frac{1}{3}, & f(b) = \frac{2}{3}, & f(c) = 0, & f(d) = 0 \\
g(a) = 0, & g(b) = 0, & g(c) = \frac{1}{3}, & g(d) = \frac{5}{6} \\
h(a) = \frac{1}{3}, & h(b) = \frac{2}{3}, & h(c) = \frac{1}{3}, & h(d) = \frac{5}{6}
\end{array}
$$

Then (X, F) is an fts where $F = \{\underline{0}, \underline{1}, f, g, h\}$. Consider the associated topology $(X, i(F))$ where $i(F) = \{\emptyset, X, \{b\}, \{d\}, \{a, b\}, \{c, d\}, \{b, d\}, \{a, b, d\}, \{b, c, d\}\}$. Clearly $(X, i(F))$ is an H-fts, but (X, F) is not an H-fts.

Theorem 2.4 [5] *Let (X, F) be a topologically generated fts. Then (X, F) is an H-fts if and only if $(X, i(F))$ is an H-fts.*

Proof Follows from theorem 1.12.   □

Theorem 2.5 [5] *Let (X, F) be an H-fts with an open fuzzy point x_λ. Then there exists $y \in X$ and $y \neq x$ such that the fuzzy point y_λ is open.*

Proof Since (X, F) is an H-fts and $x \in X$, there exists a homeomorphism θ of (X, F) such that $\theta(x) = y$ for some $y \in X$; $y \neq x$. Then $y_\lambda = \theta(x_\lambda) \in F$.   □

Theorem 2.6 [5] *Being H-fts is an additive property.*

Proof Let (X_i, F_i), $i \in \Delta$ be a family of pairwise disjoint H-fts. Let (X, F) be the sum fts of the family (X_i, F_i), $i \in \Delta$. Let $x_0 \in X$. Then $x_0 \in X_k$ for some $k \in \Delta$. Since (X_k, F_k) is an H-fts, there exists a homeomorphism θ_0 of (X_k, F_k) such that $\theta_0(x_0) \neq x_0$. Define $\theta : X \to X$ by

$$
\theta(x) = \begin{cases} \theta_0(x); & x \in X_k \\ x; & x \notin X_k \end{cases}
$$

Then θ is a homeomorphism of (X, F) such that $\theta(x_0) \neq x_0$. For, consider any $f \in F$, $\theta(f) \wedge X_k = \theta_0(f \wedge X_k) \in F_k$. For each $i \in \Delta$; $i \neq k$; $\theta(f) \wedge X_i = f \wedge X_i \in F_i$. Thus $\theta(f) \in F$, $\forall f \in F$. Similarly, $\theta^{-1}(f) \in F$, $\forall f \in F$. Since x_0 is arbitrary (X, F) is an H-fts.   □

Theorem 2.7 [5] *Being complete H-fts is an additive property.*

Proof Let (X, F) be the sum fts of a family of pairwise disjoint complete H-fts (X_i, F_i), $i \in \Delta$. Since (X_i, F_i), $i \in \Delta$ is a complete H-fts, there exists a homeomorphism θ_i on X_i such that $\theta_i(x) \neq x$, $\forall x \in X$. Define $\theta : X \to X$ by $\theta(x) = \theta_i(x)$; $x \in X_i$. Then for any $f \in F$ and for each $i \in \Delta$, $\theta(f) \wedge X_i = \theta_i(f \wedge X_i) \in F_i$ so that $\theta(f) \in F$, $\forall f \in F$. Similarly $\theta^{-1}(f) \in F$, $\forall f \in F$. Consequently θ is homeomorphism of (X, F) such that $\theta(x) \neq x$ so that (X, F) is a complete H-fts. □

Remark 2.5 Being H-fts is not a hereditary property. For example, let X be the set of real numbers. Consider $f \in I^X$ defined by

$$f(x) = \begin{cases} \frac{1}{1+|x|}; & x \in X - \{0\} \\ \frac{1}{2}; & x = 0 \end{cases}$$

Then (X, F) is an H-fts where $F = \{\underline{0}, \underline{1}, f\}$. But the subspace (Y, F_Y), where $Y = (-\infty, 0)$ is not an H-fts.

Remark 2.6 The simple extension of an H-fts need not be an H-fts. For, let (X, F) be a purely stratified fts and $a \in X$. Now consider the simple extension $(X, F(g))$ determined by the fuzzy point $g = a_\lambda$. Clearly there exists no fuzzy point $b_\lambda \in F(g)$ with $b \neq a$. Then by theorem 2.5, $(X, F(g))$ cannot be an H-fts. But the following theorem gives a situation in which the simple extension of an H-fts is again an H-fts.

Theorem 2.8 [5] *Let (X, F) be an H-fts. Then $(X, F(g))$ is an H-fts provided $g = \alpha$ for some $\alpha \in I$.*

Proof Let (X, F) be an H-fts and consider $g \in I^X$ such that $g \notin F$ and $g = \alpha$ for some $\alpha \in (0, 1)$. Since (X, F) is an H-fts for any $x \in X$, there exists a homeomorphism θ of (X, F) such that $\theta(x) \neq x$. Now consider the simple extension of (X, F) determined by g. Let $f \in F(g)$. Then $f = g_1 \vee (g_2 \wedge g)$ for some $g_1, g_2 \in F$. Let θ be any homeomorphism of (X, F). Now, $\theta^{-1}(g_1 \vee (g_2 \wedge g)) = \theta^{-1}(g_1) \vee \eta^{-1}(g_2 \wedge g)$

$$= \theta^{-1}(g_1) \vee \eta^{-1}(g_2) \wedge \eta^{-1}(g)$$
$$= \theta^{-1}(g_1) \vee \eta^{-1}(g_2) \wedge g \in F(g).$$

Also $\theta(g_1 \vee (g_2 \wedge g)) = \theta(g_1) \vee \eta(g_2) \wedge g \in F(g)$. Then θ is a homeomorphism of $(X, F(g))$ so that $(X, F(g))$ is an H-fts. $\square$

Remark 2.7 We have already seen that (X, F) being an H-fts need not imply that $(X, F(g))$; $g \in I^X$ is an H-fts. Conversely, $(X, F(g))$ being an H-fts need not imply that (X, F) is an H-fts. For example, let X be the set of all non-zero integers. Consider $f \in I^X$ defined by

$$f(x) = \begin{cases} \frac{1}{x}; & x > 0 \\ -\frac{1}{\sqrt{2}x}; & x < 0 \end{cases}$$

Then (X, F) is not an H-fts where $F = \{\underline{0}, \underline{1}, f\}$. Consider $g \in I^X$ defined by

$$g(x) = \begin{cases} \frac{1}{\sqrt{2}x}; & x > 0 \\ -\frac{1}{x}; & x < 0 \end{cases}$$

Then $\theta(x) = -x$ is a homeomorphism of $(X, F(g))$ so that $(X, F(g))$ is a complete H-fts.

Note 2.1 The inter-relations obtained so far among the various spaces considered are shown in the following diagram:

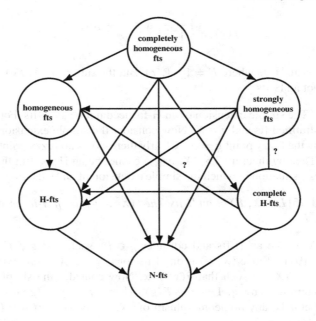

2.2 H-Fuzzy Topologies of Degree n

Generalizing the notion of H-fuzzy topological spaces, this section introduces H-fuzzy topological spaces of degree n and constructs such spaces on finite sets.

Definition 2.6 [5] An fts (X, F) is called an H-fts of degree n if $n = |Y|$ where Y is the smallest subset of X such that the subspace $(\mathscr{C}(Y), F_{\mathscr{C}(Y)})$ is an H-fts. Obviously an H-fts of degree zero is simply an H-fts.

Example 2.5 Let X be the set of all natural numbers. Consider the fuzzy subsets f, g, h of X defined by

$$f(x) = \begin{cases} \frac{2}{3}; & x \text{ is odd} \\ 0; & x \text{ is even} \end{cases}$$

$$g(x) = \begin{cases} 0; & x \text{ is odd} \\ \frac{1}{x}; & x \text{ is even} \end{cases}$$

$$h(x) = \begin{cases} \frac{2}{3}; & x \text{ is odd} \\ \frac{1}{x}; & x \text{ is even} \end{cases}$$

Then (X, F) is an fts where $F = \{\underline{0}, \underline{1}, f, g, h\}$. Clearly (X, F) is not an H-fts. Let Y be the set of all even numbers. Then the subspace $(\mathscr{C}(Y), F_{\mathscr{C}(Y)})$ is an H-fts. Consequently (X, F) is an H-fts of degree $\aleph_0$.

Remark 2.8 There are fuzzy topological spaces for which any subspace is not an H-fts. Consequently, there are fuzzy topological spaces which are not H-fts of any degree. The following example illustrates this.

Example 2.6 Let X be the set of all natural numbers. Consider $f, g \in I^X$ defined by $f(x) = \frac{1}{1+x}$ and $g(x) = \frac{1}{\sqrt{3}x}$. Let $h = f \wedge g$ and $i = f \vee g$. Then (X, F) is an fts where $F = \{\underline{0}, \underline{1}, f, g, h, i\}$. But (X, F) is not an H-fts. Further, there exists no proper subset Y of X such that the subspace $(\mathscr{C}(Y), F_{\mathscr{C}(Y)})$ is an H-fts.

Remark 2.9 All H-fts are N-fts; but this result cannot be extended to H-fts of degree n. For example, let $X = \{a, b, c\}$. Consider the fuzzy subsets f, g, h, i, j of X defined by

$$f(a) = \tfrac{1}{3}, \quad f(b) = \tfrac{1}{2}, \quad f(c) = \tfrac{1}{2}$$
$$g(a) = \tfrac{1}{5}, \quad g(b) = \tfrac{1}{5}, \quad g(c) = \tfrac{1}{4}$$
$$h(a) = \tfrac{1}{2}, \quad h(b) = \tfrac{1}{3}, \quad h(c) = \tfrac{3}{5}$$
$$i(a) = \tfrac{1}{3}, \quad i(b) = \tfrac{1}{3}, \quad i(c) = \tfrac{1}{2}$$
$$j(a) = \tfrac{1}{2}, \quad j(b) = \tfrac{1}{2}, \quad j(c) = \tfrac{3}{5}.$$

Then (X, F) is an fts where $F = \{\underline{0}, \underline{1}, f, g, h, i, j\}$. Clearly (X, F) is an H-fts of degree 1 but it is not an N-fts.

Theorem 2.9 [5] *For a given set X with $|X| = n \geq 2$, there are H-fuzzy topologies of degree $0, 1, 2, ..., n - 2$ on X.*

Proof Let $X = \{x_1, x_2, ..., x_n\}$ be the given set. Consider k such that $0 \leq k \leq n - 2$ and choose λ such that $\frac{1}{2} \leq \lambda \leq 1$. Now define a fuzzy subset f of X such that

$$f(x_i) = \begin{cases} \lambda; \ 0 \leq i \leq n - k \\ \tfrac{1}{i}; \ n - k + 1 \leq i \leq n \end{cases}$$

Then (X, F) is an fts where $F = \{\underline{0}, \underline{1}, f\}$. Clearly (X, F) is an H-fts of degree k.□

Remark 2.10 Being H-fts, is an additive property. But in general, the sum fts of a family of k pairwise disjoint H-fts of degree n need not be an H-fts of degree n. However, it is an H-fts of degree at most kn as proved by the following theorem.

Theorem 2.10 [5] *The sum fts of a family of k pair wise disjoint H-fts of degree n is an H-fts of degree at most kn.*

Proof Let $(X_i, F_i), i = 1, 2, ..., k$ be a family of pair wise disjoint H-fts of degree n..Then the subspace $(\mathscr{C}(Y_i), F_{\mathscr{C}(Y_i)})$ of (X_i, F_i) is an H-fts where Y_i is a subset of X_i with n elements. Let (X, F) be the sum fts of $(X_i, F_i), i = 1, 2, ..., k$. Now consider the subset $Y = \bigcup_{i=1}^{k} Y_i$ of X with kn elements. Then the subspace $(\mathscr{C}(Y), F_{\mathscr{C}(Y)})$ of (X, F) is the sum fts of a family of pair wise disjoint H-fts. Then by theorem 2.6 $(\mathscr{C}(Y), F_{\mathscr{C}(Y)})$ is an H-fts so that (X, F) is an H-fts of degree at most kn. □

Remark 2.11 Even though each (X_i, F_i), $i \in \Delta$ is not an H-fts of any degree, the sum fts may be an H-fts. For example, let $\mathbb{N}$ be the set of all natural numbers and for each $i \in \mathbb{N}$, define $X_i = \{x : x \in \mathbb{N}, 10(i-1) < x \le 10(i)\}$. Consider the fuzzy subset f_i of X_i defined by $f_i(x) = \frac{1}{x-10(i-1)}$. Then (X_i, F_i), $i \in \mathbb{N}$ is a family of pair wise disjoint fts where $F_i = \{\underline{0}, \underline{1}, f_i\}$. Clearly for each $i \in \mathbb{N}$, (X_i, F_i) is not an H-fts of any degree. Consider the sum fts (X, F) and $x \in X$. Then $x \in X_k$ for some $k \in \mathbb{N}$.

Define $\theta : X \to X$ by

$$\theta(x) = \begin{cases} x + 10; & k \text{ is odd} \\ x - 10; & k \text{ is even} \end{cases}$$

Clearly the homeomorphism θ makes (X, F) an H-fts.

Remark 2.12 It can be easily verified that the simple extension of an H-fts of degree n need not be an H-fts of any degree. If it happens to be an H-fts of some degree k, then k can take values independent of n.

2.3 Exercises

2.1 *Show that:*

1. *a strongly homogeneous fts is always a homogeneous fts.*
2. *a homogeneous fts is always an N-fts.*
3. *every completely homogeneous fts is strongly homogeneous.*

2.2 *Show that the converses of (a), (b) and (c) of the exercise 2.1 are not true.*

2.3 *Show that a purely stratified fts is always a complete H-fts.*

2.4 *Let (X, F) be an N-fts, then show that $(X, i(F))$ is also an N-fts.*

2.5 *Let (X, F) be a topologically generated fts. Then prove that (X, F) is an N-fts if and only if $(X, i(F))$ is an N-fts.*

2.6 *Show that being N-fts is an additive property.*

2.7 *Show that being N-fts is not a hereditary property.*

2.8 *Show that the simple extension of an N-fts need not be an N-fts.*

2.9 *Let (X, F) be an N-fts. Then show that $(X, F(g))$ is an N-fts provided $g = \underline{\alpha}$ for some $\alpha \in I$.*

2.10 *Show that a purely stratified fts is always a completely homogeneous fts.*

2.11 *For a homogeneous fts* (X, F), *show that* $X_\alpha \neq \emptyset$ *for some* $\alpha \in [0, 1]$ *implies* $X_\alpha = X$.

2.12 *Let* (X, T) *be a topological space. Show that* (X, T) *is strongly homogeneous if and only if* $(X, \omega(T))$ *is strongly homogeneous.*

2.13 *Let* (X, F) *be a strongly homogeneous fts. Then show that* $(X, i(F))$ *is also strongly homogeneous.*

2.14 *Let* (X, F) *be a topologically generated fts. Show that* (X, F) *is strongly homogeneous if and only if* $(X, i(F))$ *is strongly homogeneous.*

2.15 *Show that being completely homogeneous is a hereditary property.*

2.16 *Show that being strongly homogeneous is not a hereditary property.*

2.17 *Show that the simple extension of a strongly homogeneous fts need not be strongly homogeneous.*

2.18 *Let* (X, F) *be a strongly homogeneous fts. Then* $(X, F(g))$ *is strongly homogeneous provided* $g = \underline{\alpha}$ *for some* $\alpha \in I$.

References

1. Mathew, S.C., Jose, A.: Nice fuzzy topological spaces. STARS. Int. J. (Sci.) **1**(2), 156–162 (2007)
2. Jose, V.K., Johnson, T.P.: Complete homogeneity and reversibility in L-topology. Far East J. Math. Sci. **16**(3), 357–362 (2005)
3. Jose, A., Mathew, S.C.: Strongly homogeneous fuzzy topological spaces. Int. J. Fuzzy Math. Syst. **1**(3), 263–271 (2011)
4. Johnson, T.P.: Completely homogeneous fuzzy topological spaces. J. Fuzzy Math. **1**(3), 495–500 (1993)
5. Mathew, S.C., Jose, A.: H-fuzzy topological spaces. J. Fuzzy Math. **18**(1), 117–128 (2010)

Chapter 3
Invertible Fuzzy Topological Spaces

The chapter focuses on the basic nature of invertible fuzzy topological space with special reference to the role of fuzzy points. Certain conditions necessary for an inverting fuzzy subset in the respective spaces are derived. In addition, situations under which a given fuzzy set is not an inverting fuzzy set are explored. Completely invertible fuzzy topological spaces are also defined and some characterizations are obtained. The basic properties of such spaces are examined thoroughly. Since home-omorphisms play a vital role in the invertibility of fuzzy topological space, the relationship of invertible and completely invertible fts with H-fts, N-fts and homogeneous fts is investigated. The orbits in invertible fuzzy topological spaces are also studied and described in the best possible manner.

3.1 Invertibility of Fuzzy Topological Spaces

This section includes some basic properties of invertible fuzzy topological spaces.

Definition 3.1 [1] An open subset g of an fts (X, F) is called an inverting set if there is a homeomorphism θ of (X, F) such that $\theta(\mathcal{C}(g)) \leq g$. This homeomorphism θ is called an inverting map for g. An fts (X, F) is said to be invertible if it has an inverting proper subset.

When we say (X, F) is invertible with respect to g, it is understood that $g \in F$ and $g \neq \underline{1}$.
We begin with the following observation:

Theorem 3.1 [2] *Let (X, F) be an fts, invertible with respect to g and let $h \in F$ be such that $h \geq g$. Then (X, F) is invertible with respect to h also. Moreover there is a common inverting map for both g and h.*

A. Jose and S. C. Mathew, *Invertible Fuzzy Topological Spaces*,
https://doi.org/10.1007/978-981-19-3689-0_3

Proof Let θ be an inverting homeomorphism for g. Then $\theta(\mathcal{C}(g)) \leq g \Rightarrow \mathcal{C}(g) \leq \theta^{-1}(g) \Rightarrow \mathcal{C}(h) \leq \theta^{-1}(g) \Rightarrow \mathcal{C}(h) \leq \theta^{-1}(h) \Rightarrow \theta(\mathcal{C}(h)) \leq h$. Hence (X, F) is invertible with respect to h and θ is an inverting map for h also. $\square$

Following theorems pinpoint certain conditions necessary for an inverting fuzzy subset in the respective spaces.

Theorem 3.2 [2] *If (X, F) is an fts invertible with respect to g, then $|supp\,\mathcal{C}(g)| \leq |supp\,g|$.*

Proof Since g is an inverting fuzzy subset of (X, F), there exists a homeomorphism θ of (X, F) such that $\theta(\mathcal{C}(g)) \leq g$
$$\Rightarrow |supp\,\theta(\mathcal{C}(g))| \leq |supp\,g| \Rightarrow |supp\,(\mathcal{C}(g))| \leq |supp\,g|. \qquad \square$$

Theorem 3.3 [2] *Let (X, F) be an fts invertible with respect to g where X is finite. Then $|ss\,\mathcal{C}(g)| \leq \frac{|X|}{2}$.*

Proof (X, F) is invertible with respect to g
  $\Rightarrow$ there exists a homeomorphism
θ of (X, F) such that $\theta(\mathcal{C}(g)) \leq g$
  $\Rightarrow |ss\,\mathcal{C}(g)| \leq |ssg|$
  $\Rightarrow |ss\,\mathcal{C}(g)| \leq |X| - |ws\,g|$
  $\Rightarrow |ss\,\mathcal{C}(g)| \leq |X| - |ss\,\mathcal{C}(g)|$
  $\Rightarrow 2|ss\,\mathcal{C}(g)| \leq |X|$. $\square$

Theorem 3.4 [2] *Let (X, F) be an fts invertible with respect to g where X is finite. Then $|X| \leq 2|supp\,g|$.*

Proof By theorem 3.3, $|ss\,\mathcal{C}(g)| \leq \frac{|X|}{2}$
  $\Rightarrow |X| - |ws\,\mathcal{C}(g)| \leq \frac{|X|}{2}$
  $\Rightarrow \frac{|X|}{2} \leq |ws\,\mathcal{C}(g)|$.
Since $|supp\,g| \geq |ws\,\mathcal{C}(g)|$, the result follows. $\square$

Theorem 3.5 [2] *Let (X, F) be an fts and let $g \in F$ and $g \neq \frac{1}{2}$ contains no strong fuzzy points. Then (X, F) is not invertible with respect to g.*

Proof Since $g \neq \frac{1}{2}$ and contains no strong fuzzy points, $\mathcal{C}(g)$ contains at least one strong fuzzy point. Then there exists no homeomorphism θ of (X, F) such that $\theta(\mathcal{C}(g)) \leq g$. Consequently (X, F) is not invertible with respect to g. $\square$

Remark 3.1 From the proof of the above theorem, it follows that any non-constant fuzzy subset containing only weak fuzzy points cannot be an inverting fuzzy subset for an fts.

Theorem 3.6 [2] *Let (X, F) be an fts and $g \in F$ be such that supp $g \neq X$. If g doesn't contain a crisp singleton, then g is not an inverting fuzzy subset for (X, F).*

Proof Since $supp\ g \neq X$ and g doesn't contain a crisp singleton, $\mathcal{C}(g)$ contains a crisp singleton so that there exists no homeomorphism θ of (X, F) such that $\theta(\mathcal{C}(g)) \leq g$. □

Theorem 3.7 [2] *Let (X, F) be an fts in which no weak fuzzy subset is closed and $|X_{\frac{1}{2}}| = 0$. If $g \in F$ is such that $g \wedge \mathcal{C}(g)$ is a weak fuzzy subset then (X, F) is not invertible with respect to g.*

Proof Since $g \wedge \mathcal{C}(g)$ is a weak fuzzy subset, g doesn't contain any crisp singleton. Also $g(x) < \frac{1}{2}$ for some $x \in X$. For, otherwise $\mathcal{C}(g)$ is a weak closed fuzzy subset of (X, F), a contradiction. Now if possible let (X, F) be invertible with respect to g and θ be an inverting map for g. Then $\theta(g)(x) > \frac{1}{2}$ for all $x \in X$ for which $g(x) < \frac{1}{2}$. Take $f = g \vee \theta(g)$ so that $f(x) \geq \frac{1}{2}$, $\forall x \in X$ and f doesn't contain any crisp singleton. Then $\mathcal{C}(f)$ is a closed weak fuzzy subset of (X, F), a contradiction. □

Now we identify certain classes of fuzzy topological spaces that are invertible.

1. An fts with a closed weak fuzzy point.
2. A quasi-separated fts.
3. An H-fts containing a well-closed fuzzy point.

Theorem 3.8 [2] *Every fts can be embedded in an invertible fts.*

Proof Let (X, F) be an fts. If (X, F) is invertible, there is nothing to prove. Suppose (X, F) is not invertible, then by theorem 3.2, the weak fuzzy points in (X, F) are not closed. Let x_λ be a weak fuzzy point in (X, F). Then the simple extension $(X, F(\mathcal{C}(x_\lambda)))$ is invertible. □

3.2 Completely Invertible Fuzzy Topological Spaces

This section studies fuzzy topological spaces in which every proper open fuzzy subset is an inverting fuzzy subset. Such spaces are called completely invertible.

Definition 3.2 [2] A non-trivial fts (X, F) is said to be completely invertible if for every $g \in F$ and $g \neq \underline{0}, \underline{1}$ there is a homeomorphism θ of (X, F) such that $\theta(\mathcal{C}(g)) \leq g$.

It should be noted that for a completely invertible fts every proper open fuzzy subset is an inverting fuzzy set.

Even if it is a matter of "complete" invertibility forget everything but base elements!

Theorem 3.9 [2] *Let (X, F) be an fts with $\mathcal{B}$ as a base. Then (X, F) is completely invertible if and only if (X, F) is invertible with respect to all members of $\mathcal{B}$.*

Proof The necessary part is obvious. Conversely suppose (X, F) is invertible with respect to b for all $b \in \mathcal{B}$. By theorem 1.3, any $f \in F$ and $f \neq \underline{0}$ is either a base element or there exists a $b \in \mathcal{B}$ such that $b < f$. Consequently by theorem 3.1, (X, F) is completely invertible. $\qquad\square$

Theorem 3.10 [2] *An fts (X, F) is completely invertible if and only if for each proper closed fuzzy subset h and each $g \in F$ and $g \neq \underline{0}$, there is a homeomorphism θ of (X, F) such that $\theta(h) \leq g$.*

Proof The sufficient part is obvious. Assume that (X, F) is completely invertible. Let g be a non-empty open fuzzy subset and h be a proper closed fuzzy subset of X. Then the following two cases arise:

Case 1: $g \wedge \mathcal{C}(h) \neq \underline{0}$. Let θ be an inverting homeomorphism for $g \wedge \mathcal{C}(h)$, then we have $\theta(h) \leq g$.

Case 2: $g \wedge \mathcal{C}(h) = \underline{0}$. Then $\mathcal{C}(h) \leq \mathcal{C}(g)$. Let θ be an inverting homeomorphism for $\mathcal{C}(h)$ and η be an inverting homeomorphism for g. Then we have $\theta(h) \leq \mathcal{C}(h)$ implies $\eta(\theta(h)) \leq g$. Since $\eta \circ \theta$ is a homeomorphism, the result follows. $\qquad\square$

A completely invertible finite c-fts is completely characterized below:

Theorem 3.11 [2] *A finite c-fts (X, F) is completely invertible if and only if $|X|$ is even and $F = \{\underline{0}, \underline{1}, f, \mathcal{C}(f)\}$.*

Proof Suppose that (X, F) is completely invertible. Let f be an inverting fuzzy subset of (X, F) and θ be an inverting map of f. Then by theorem 3.4, $|supp\, f| \geq \frac{|X|}{2}$. Let $g_0 = f$ and $\theta_0 = \theta$. Define $g_1 = g_0 \wedge \theta_0(g_0)$. Clearly $g_1 \in F$ and $|supp\, g_1| < |supp\, g_0|$. Let θ_1 be an inverting map of g_1. Let $g_2 = g_1 \wedge \theta_1(g_1)$. Clearly $g_2 \in F$ and $|supp\, g_2| < |supp\, g_1|$. Continuing like this, after a certain stage there exists a $g_n \in F$ and $g_n \neq \underline{0}$, where $n \in \mathbb{N}$ and $g_n = g_{n-1} \wedge \theta_{n-1}(g_{n-1})$ such that $|supp\, g_n| \leq \frac{|X|}{2}$. Since (X, F) is completely invertible, $|supp\, g_n| = \frac{|X|}{2}$. Now consider $k = \theta_n(g_n) \wedge f \in F$. Then $|supp\, k| < \frac{|X|}{2}$ so that by theorem 3.4, $k = \underline{0}$. Since $\theta_n(g_n) = \mathcal{C}(g_n)$, $k = \underline{0} \Rightarrow g_n = f$. Hence $|supp\, f| = \frac{|X|}{2}$. Thus $f \in F$ and $f \neq \underline{0} \Rightarrow |supp\, f| = \frac{|X|}{2}$.

Now if possible let $g \in F$ and $g \neq f$. Then from above we have $|supp\, g| = \frac{|X|}{2}$. Let $h = g \wedge f$, then $h \in F$ and $|supp\, h| < \frac{|X|}{2}$. Since (X, F) is completely invertible, $h = \underline{0}$ which implies that $g = \mathcal{C}(f)$. Hence $F = \{\underline{0}, \underline{1}, f, \mathcal{C}(f)\}$.

Conversely assume that $|X|$ is even and $F = \{\underline{0}, \underline{1}, f, \mathcal{C}(f)\}$. Let $supp\, f = \{x_1, x_2, ..., x_n\}$ and $supp\, \mathcal{C}(f) = \{y_1, y_2, ..., y_n\}$. Define $\theta : X \to X$ by $\theta(x_i) = y_i$ and $\theta(y_i) = x_i$ where $i = 1, 2, ..., n$. Clearly θ is an inverting map for f and $\mathcal{C}(f)$ in (X, F). Consequently (X, F) is completely invertible. $\qquad\square$

The following theorem gives a sufficient condition for a fuzzy topological space to be completely invertible.

Theorem 3.12 [2] *Let (X, F) be an fts invertible with respect to $g = \bigwedge\limits_{f \in F - \{\underline{0}\}} f$.*

Then (X, F) is completely invertible.

Proof Let (X, F) be an fts invertible with respect to $g = \bigwedge\limits_{f \in F - \{0\}} f$. Clearly $g \leq f$ for all $f \in F$; $f \neq \underline{0}$ and then by theorem 3.1, (X, F) is invertible with respect to every $f \in F$ and $f \neq \underline{0}$.   □

Now we discuss certain properties of completely invertible fuzzy topological spaces.

Theorem 3.13 [2] *Let (X, F) be a completely invertible fts, where X is finite. Then $|X_\alpha| = 0$ for all $\alpha \in (0, \frac{1}{2})$.*

Proof Suppose (X, F) is completely invertible. Assume that there exists an open subset g_1 such that $g_1(x) = \alpha$ for some x in X and $\alpha \in (0, \frac{1}{2})$. Then there exists a homeomorphism θ_1 of (X, F) such that $\theta_1^{-1}(x) = y$; $y \neq x$. Let $g_2 = g_1 \wedge \theta_1^{-1}(g_1)$. Clearly $|ss\,\mathcal{C}(g_2)| \geq 2$. Let θ_2 be an inverting homeomorphism for g_2. Now let $g_3 = g_2 \wedge \theta_2^{-1}(g_2)$. Then $|ss\,\mathcal{C}(g_3)| \geq 4$. Proceeding like this, after a certain stage we get a $g_i \in F$ such that $|ss\,\mathcal{C}(g_i)| > \frac{|X|}{2}$. But by theorem 3.3, this is a contradiction to the fact that (X, F) is completely invertible. Consequently $|X_\alpha| = 0$ for all $\alpha \in (0, \frac{1}{2})$.   □

Corollary 3.1 [2] *Let (X, F) be a completely invertible fts and $|X_\alpha| > 0$, for some $\alpha \in (0, \frac{1}{2})$. Then X is infinite.*

Theorem 3.14 [2] *Let (X, F) be a completely invertible fts and $|X_\alpha| > 0$ for some $\alpha \in [0, \frac{1}{2})$. Then (X, F) is an N-fts.*

Proof Let $\beta \in [0, \frac{1}{2})$ such that X_β is non-empty and let $x_0 \in X_\beta$. Then there exists an open fuzzy subset g in X such that $g(x_0) = \beta$. Assume that (X, F) is not an N-fts. Then identity is the inverting map for every $f \in F$. But $\mathcal{C}(g) \not\leq g$, so that (X, F) is not invertible with respect to g, which is a contradiction. Consequently (X, F) is an N-fts.   □

Corollary 3.2 [2] *Let (X, F) be a completely invertible fts containing a non-empty nearly crisp proper fuzzy subset. Then (X, F) is an N-fts.*

Proof Let g be a non-empty nearly crisp proper fuzzy subset in (X, F). Then $\mathcal{C}(\overline{g})(x) = 0$ for some $x \in X$ so that $|X_0| > 0$. Now the corollary follows from the above theorem.   □

Remark 3.2 Let (X, F) be a completely invertible fts which is not an N-fts, then $|X_\alpha| = 0$ for all $\alpha \in [0, \frac{1}{2})$. Also in this case (X, F) cannot contain a nearly crisp proper fuzzy subset.

Theorem 3.15 [2] *Let (X, F) be a completely invertible fts and $\bigcup\limits_{0 \leq \alpha < \frac{1}{2}} X_\alpha = X$. Then (X, F) is an H-fts.*

Proof Since $\bigcup\limits_{0\leq\alpha<\frac{1}{2}} X_\alpha = X$, then for each $x \in X$ there exists some open fuzzy subset f_x in X such that $f_x(x) \in [0, \frac{1}{2})$. Since (X, F) is completely invertible, then for each x in X there exist a homeomorphism θ of (X, F) such that $\theta(x) \neq x$. For, otherwise let there be an $x \in X$ such that $\theta(x) = x$ for every homeomorphism θ of (X, F). Then $\theta(\mathscr{C}(f_x)) \not\leq f_x$ for every homeomorphism θ of (X, F) so that (X, F) is not invertible with respect to f_x. This contradiction proves the claim and hence (X, F) is an H-fts.   □

3.3 Homogeneity and Invertibility

Invertibility and homogeneity are independent and in certain situations one implies the other. Exploration of such situations is highlighted here.

Remark 3.3 A homogeneous fts need not be invertible. For example, let $X = \{a, b, c, d\}$. Consider $f, g \in I^X$ defined by

$$f(a) = \tfrac{2}{7}, \quad f(b) = \tfrac{2}{9}, \quad f(c) = \tfrac{2}{9}, \quad f(d) = \tfrac{2}{7}$$
$$g(a) = \tfrac{2}{9}, \quad g(b) = \tfrac{2}{7}, \quad g(c) = \tfrac{2}{7}, \quad g(d) = \tfrac{2}{9}.$$

Then (X, F) is an fts where $F = \{\underline{0}, \underline{1}, \tfrac{2}{7}, \tfrac{2}{9}, f, g\}$. Now the mappings θ_1, θ_2, θ_3 : $X \to X$, defined by

$$\theta_1(a) = b, \quad \theta_1(b) = a, \quad \theta_1(c) = d, \quad \theta_1(d) = c$$
$$\theta_2(a) = c, \quad \theta_2(b) = d, \quad \theta_2(c) = a, \quad \theta_2(d) = b$$
$$\theta_3(a) = d, \quad \theta_3(b) = c, \quad \theta_3(c) = b, \quad \theta_3(d) = a$$

are homeomorphisms of (X, F). Thus (X, F) is strongly homogeneous. Since any proper open fuzzy subset of (X, F) contains no strong fuzzy point, (X, F) is not invertible by theorem 3.5.

Homogeneous fts becomes invertible under certain conditions:

Theorem 3.16 [3] *If a finite fts (X, F) is homogeneous and $X_\beta \neq \emptyset$ for some $\beta \in [\frac{1}{2}, 1]$, then (X, F) is invertible.*

Proof Let (X, F) be a homogeneous fts and $X_\beta \neq \emptyset$ for some $\beta \in [\frac{1}{2}, 1]$. Then by exercise 2.11 $X_\beta = X$ so that for each $x \in X$ there exists an open fuzzy subset h_x such that $h_x(x) = \beta$. Now there arise two cases:

Case 1: $\beta \neq 1$. Now for each $x \in X$, choose one open fuzzy subset, say g_x such that $g_x(x) = \beta$ and let $g = \bigvee\limits_{x \in X} g_x$. Clearly $g \neq \underline{1}$ and $g \geq \underline{\beta}$. Thus (g, e) is an inverting pair of (X, F).

Case 2: $\beta = 1$. Then for each $x \in X$, there exists open fuzzy subset g_x such that $g_x(x) = 1$. Now choose an $g \neq \underline{1}$ in F such that $|S_1(g)|$ is maximum. Clearly $|S_1(g)| \geq \frac{|X|}{2}$ and since (X, F) is homogeneous there exists a homeomorphism θ of (X, F) such that $\theta(\mathcal{C}(g)) \leq g$.

Consequently (X, F) is invertible.   □

Converse of the above theorem is not true as shown below.

Example 3.1 Let X be the set of all non-negative numbers. Consider $h \in I^X$ defined by

$$h(x) = \begin{cases} 1; & x = 0 \\ \frac{x}{x+1}; & \text{otherwise} \end{cases}$$

Then (X, F) is an fts where $F = \{\underline{0}, \underline{1}, h\}$. Here (X, F) is not homogeneous, but it is invertible.

Remark 3.4 Even complete invertibility need not imply homogeneity. For example, let X be the set of all natural numbers and let $Y = \{4x : x \in X\}$. For each $k \in Y$, define $f_k, g_k \in I^X$ by

$$f_k(x) = \begin{cases} \frac{1}{3}; & 1 \leq x \leq k \\ \frac{2}{3}; & \text{otherwise} \end{cases}$$

$$g_k(x) = \begin{cases} \frac{1}{3}; & k < x \leq 2k \\ \frac{2}{3}; & \text{otherwise} \end{cases}$$

Let F be the fuzzy topology on X generated by $\{f_k, g_k : k \in Y\}$. Now for each $k \in Y$, consider $\theta_k : X \to X$ defined by

$$\theta_k(x) = \begin{cases} x + k; & 1 \leq x \leq k \\ x - k; & k < x \leq 2k \\ x; & \text{otherwise.} \end{cases}$$

Clearly θ_k is a homeomorphism of (X, F). Also for each $k \in Y$, θ_k is an inverting map for both f_k and g_k so that (X, F) is completely invertible. Here $2, 7 \in X$, but there exist no homeomorphism θ of (X, F) such that $\theta(2) = 7$. Hence (X, F) is not homogeneous.

In contrast, from the characterization of a completely invertible finite c-fts, the following results are evident for a completely invertible c-fts (X, F):

- (X, F) is always strongly homogeneous.
- If $|X| = 2$, then (X, F) is completely homogeneous.
- if $|X| > 2$, then (X, F) cannot be completely homogeneous.

If the inverting subset as a subspace is homogeneous, then the parent fuzzy topological space is also homogeneous.

Theorem 3.17 [3] *Let (X, F) be an fts invertible with respect to a crisp subset A. If (A, F_A) is homogeneous, then (X, F) is also homogeneous.*

Proof Let θ be an inverting homeomorphism for A. Then $\theta^{-1}(A) \geq \mathcal{C}(A)$. Let (A, F_A) be homogeneous. Then $(\theta^{-1}(A), F_{\theta^{-1}(A)})$ is also homogeneous. Let $x, y \in X$. Then the following three cases arise:

Case 1: $x, y \in A$. Since A is homogeneous, there exists a homeomorphism η_A of (A, F_A) such that $\eta_A(x) = y$. Define $\eta : X \to X$ by

$$\eta(x) = \begin{cases} \eta_A(x); & x \in A \\ x; & \text{otherwise} \end{cases}$$

Clearly η is a homeomorphism of (X, F) such that $\eta(x) = y$.

Case 2: $x, y \in \theta^{-1}(A)$. Since $(\theta^{-1}(A), F_{\theta^{-1}(A)})$ is homogeneous, this is similar to that of case 1.

Case 3: Exactly one of $x, y \in A$. Let $x \in A$ and $y \notin A$. Since θ is an inverting map for A, $\theta(y) \in A$. Since A is homogeneous, there exists a homeomorphism η_A of (A, F_A) such that $\eta_A(x) = \theta(y)$. Define $\eta : X \to X$ by

$$\eta(x) = \begin{cases} \eta_A(x); & x \in A \\ x; & \text{otherwise} \end{cases}$$

Now $\theta^{-1} \circ \eta$ is a homeomorphism of (X, F) such that $\theta^{-1} \circ \eta(x) = \theta^{-1}(\eta(x)) = y$. Consequently (X, F) is homogeneous.   □

If (X, F) is not invertible with respect to B, then it need not be homogeneous, even if (B, F_B) is strongly homogeneous as shown in the following example. Hence, invertibility is indispensable for the above theorems.

Example 3.2 Let X be the set of all real numbers and B be the set of all natural numbers. Consider $f, g, h \in I^X$ defined by

$$f(x) = \begin{cases} \frac{1}{2}; & x \in B \\ |x|; & -1 < x < 1 \\ \frac{1}{|x|}; & \text{otherwise} \end{cases}$$

$$g(x) = \begin{cases} 1; & x \in B \\ |x|; & -1 < x < 1 \\ \frac{1}{|x|}; & \text{otherwise} \end{cases}$$

$$h(x) = \begin{cases} \frac{1}{2}; & x \in B \\ 0; & \text{otherwise} \end{cases}$$

Then (X, F) is an fts where $F = \{\underline{0}, \underline{1}, f, g, h, B\}$ and B is not a inverting fuzzy subset of (X, F). Clearly (B, F_B) is strongly homogeneous, but (X, F) is not even homogeneous.

Remark 3.5 In contrast to homogeneity and strong homogeneity, complete homogeneity cannot be transferred from the subspace to parent space with the help of invertibility. For example, let X be the set of all non-zero real numbers. For each $\beta \in [\frac{1}{2}, 1]$, define f_β, $g_\beta \in I^X$ by

$$f_\beta(x) = \begin{cases} \beta; & x < 0 \\ 1 - \beta; & x > 0 \end{cases}$$

$$g_\beta(x) = \begin{cases} 1 - \beta; & x < 0 \\ \beta; & x > 0 \end{cases}$$

Let F be the fuzzy topology on X generated by $G = \{f_\beta, g_\beta \in I^X : \beta \in [\frac{1}{2}, 1]\}$. Let $B = f_1$. Then $B \in F$ and (X, F) is invertible with respect to B and (B, F_B) is completely homogeneous, but (X, F) is not completely homogeneous.

3.4 Orbits in Invertible Fuzzy Topological Spaces

This section elaborates the properties of orbits in invertible fuzzy topological space.

Definition 3.3 [3] If (X, F) is an fts and $x \in X$, then the orbit of x, O_x, is the set of all images of x under elements of $G(F)$.

O_x is a homogeneous subspace and if (X, F) is homogeneous, then $O_x = X$. Thus (X, F) is homogeneous if and only if $O_x = X$ for every $x \in X$. If O_x and O_y are distinct orbits in X, then $O_x \cap O_y = \emptyset$ and $\bigcup_{x \in X} O_x = X$. Thus orbits are equivalence classes of X.

We begin with the following beautiful observation:

Theorem 3.18 [3] *The orbits in a completely invertible fts are dense.*

Proof Let (X, F) be a completely invertible fts and let $f \neq \underline{0}$ be an open subset of (X, F). Let $x \in X$. If $(f \wedge O_x)(x) = 0$, then $f(x) = 0$. Since (X, F) is completely invertible, there exists a homeomorphism θ of (X, F) such that $\theta \mathcal{C}(f) \leq f$. Hence there exists a $y \in X$ such that $f(y) = 1$ and $\theta(y) = x$. Since $y \in O_x$, $O_x \wedge f \neq \underline{0}$. Thus for the orbit O_x of some $x \in X$, $f \in F$ and $f \neq \underline{0} \Rightarrow f \wedge O_x \neq \underline{0}$. Now if possible assume that $\overline{O_x} = A \neq \underline{1}$. Then $B = \underline{1} \wedge \mathcal{C}(A) \in F$ and hence $B \wedge O_x = \underline{0}$, a contradiction. Consequently $\overline{O_x} = \underline{1}$ so that O_x is a dense fuzzy subset of (X, F). $\qquad\qquad\qquad\square$

The following example shows that the orbits in an invertible fts need not be dense.

Example 3.3 Let X be the set of all natural numbers. Now consider $p, q, r, s \in I^X$ defined by

$$p(x) = \frac{x}{x+1}, \quad \forall x \in X$$

$$q(x) = \begin{cases} 0; & x \text{ is odd} \\ 1; & x \text{ is even} \end{cases}$$

$$r(x) = \begin{cases} 0; & x \text{ is odd} \\ \frac{x}{x+1}; & x \text{ is even} \end{cases}$$

$$s(x) = \begin{cases} \frac{x}{x+1}; & x \text{ is odd} \\ 1; & x \text{ is even} \end{cases}$$

Then (X, F) is an fts where $F = \{\underline{0}, \underline{1}, p, q, r, s\}$. Clearly $G(F) = \{e\}$ and e is an inverting map for p in (X, F). Here $O_x = \{x\}$ for all $x \in X$ and $\overline{O_1} = \mathcal{C}(q) \neq \underline{1}$. Thus (X, F) is not homogeneous.

Theorem 3.19 [3] *The orbits in a completely invertible non-homogeneous fts are neither open nor closed.*

Proof Let (X, F) be a completely invertible and non-homogeneous fts. Since (X, F) is not homogeneous, $O_x \neq \underline{1}$. If possible assume that $\overline{O}_x = O_x$. Then by theorem 3.18, $O_x = \underline{1}$, a contradiction. Consequently O_x is not closed. Now if possible assume that O_x is open for some $x \in X$. If $y \in X$ and $y \notin O_x$, then $O_y \wedge O_x = \underline{0}$ so that $O_y \leq \mathcal{C}(O_x)$. Hence $\overline{O_y} \leq \mathcal{C}(O_x) < \underline{1}$, a contradiction by theorem 3.18. $\square$

Theorem 3.20 [3] *If (X, F) and (Y, G) are completely invertible fuzzy topological spaces which are homeomorphic, then the orbits in (Y, G) are homeomorphic images of orbits in (X, F).*

Proof Let $y \in Y$ and O_y denote the orbit of y in Y. Let $\theta : X \to Y$ be a homeomorphism and $\theta^{-1}(y) = x$. Let O_x denote the orbit of $x \in X$. It must be proved that $O_y = \theta(O_x)$. Let $z \in \theta(O_x)$. Then $\theta^{-1}(z) \in O_x \Rightarrow \exists \phi \in G(F)$ such that $\phi(x) = \theta^{-1}(z)$. Thus $z = \theta(\phi(x)) = \theta(\phi(\theta^{-1}(y)))$ so that $z \in O_y$. Hence $\theta(O_x) \subset O_y$. Conversely, let $z \in O_y$. Then there is a homeomorphism η of (Y, G) such that $\eta(y) = z$. Consider $\theta^{-1}(\eta(\theta(x)) = \theta^{-1}(\eta(y)) = \theta^{-1}(z) \Rightarrow \theta^{-1}(z) \in O_x \Rightarrow z \in \theta(O_x)$. Thus $O_y \subset \theta(O_x)$. Consequently $O_y = \theta(O_x)$. $\square$

The following theorem characterizes orbits in an invertible fts.

Theorem 3.21 [3] *A non-empty subset O of an invertible fts (X, F) is an orbit in (X, F) if and only if (O, F_O) is homogeneous and invariant under homeomorphisms of (X, F).*

Proof Suppose that O is an orbit in (X, F). Then clearly (O, F_O) is homogeneous and invariant under homeomorphisms of (X, F). Conversely, (O, F_O) is homogeneous and invariant under homeomorphisms of (X, F) and let $x \in O$. It is to be proved that $O = O_x$. If $y \in O$, then there exists a homeomorphism θ of (X, F) such that $\theta(x) = y$ so that $y \in O_x$. Conversely if $y \in O_x$, then there exists a homeomorphism η of (X, F) such that $\eta(x) = y$. But since O is invariant under homeomorphism, $y \in O$. Consequently $O = O_x$. $\square$

Theorem 3.22 [3] *Let B be any finite union of orbits in a completely invertible (X, F). If the subspace (B, F_B) is non-trivial, then it is completely invertible.*

Proof Let $g \in F_B$ and $g \neq \underline{0}, \underline{1}$. Then there exists an $f \in F$; $f \neq \underline{1}$ such that $f \wedge B = g$. Clearly $f(x) = g(x)$ for all $x \in B$. Since (X, F) is completely invertible, there exists a homeomorphism θ of (X, F) such that $\theta(\mathcal{C}(f)) \leq f \Rightarrow \theta(\mathcal{C}(f)) \wedge B \leq f \wedge B$. Since B is the union of orbits, the restriction of θ to B, $\theta|_B$ is a homeomorphism of (B, F_B). Also $\theta(\mathcal{C}(f)) \wedge B = \theta|_B(\mathcal{C}(f)) = \theta|_B(\mathcal{C}(g))$. Consequently $\theta|_B(\mathcal{C}(g)) \leq g$ so that $\theta|_B$ is an inverting map for g in (B, F_B). $\qquad\square$

As an immediate consequence, it is obtained that if an orbit as a subspace is non-trivial, then that subspace is completely invertible.

Remark 3.6 From the above properties of orbits, it follows that every completely invertible fuzzy topological space is the disjoint union of homogeneous dense subspaces in which every non-trivial subspace is completely invertible.

Orbit of an invertible fts need not be invertible as a subspace as shown in the following example.

Example 3.4 Let $X = \{u, v, w, x\}$. Consider $f, g, h \in I^X$ defined by

$$f(u) = 1, \quad f(v) = 1, \quad f(w) = \tfrac{2}{5}, \quad f(x) = 1$$
$$g(u) = 1, \quad g(v) = 1, \quad g(w) = 1, \quad g(x) = \tfrac{2}{5}$$
$$h(u) = 1, \quad h(v) = 1, \quad h(w) = \tfrac{2}{5}, \quad h(x) = \tfrac{2}{5}$$

Then (X, F) is an fts where $F = \{\underline{0}, \underline{1}, \tfrac{2}{5}, f, g, h\}$. Clearly $\theta : X \to X$ defined by $\theta(u) = v$, $\theta(v) = u$, $\theta(w) = x$, $\theta(x) = w$ is an inverting map for f. Here $O_u = \{u, v\}$ and (O_u, F_{O_u}) is not invertible.

An orbit need not be an inverting subset for a fuzzy topological space. The next theorem portrays that in an invertible fts, there is always an orbit which is invertible as a subspace.

Theorem 3.23 [3] *Every invertible fts has an orbit which is invertible as a subspace.*

Proof If (X, F) is homogeneous, then $O_x = X$ for all $x \in X$ so that there is nothing to prove. Otherwise, let f be an inverting fuzzy subset of (X, F) and θ be an inverting map for f. Then $f(x) < 1$ for some $x \in X$. Now let $A = O_x$ and $g = f \wedge A$. Then $g \in F_A$ with $g \neq \underline{1}_A$ and $g(x) = f(x)$ for all $x \in A$. Also $\theta(\mathcal{C}(f)) \leq f \Rightarrow \theta(\mathcal{C}(f)) \wedge A \leq f \wedge A$. Since A is an orbit, the restriction of θ to A, $\theta|_A$ is a homeomorphism of (A, F_A). Since $\theta(\mathcal{C}(f)) \wedge A = \theta|_A(\mathcal{C}(f)) = \theta|_A(\mathcal{C}(g))$, we have $\theta|_A(\mathcal{C}(g)) \leq g$. Consequently $\theta|_A$ is an inverting map for g in (A, F_A). $\qquad\square$

3.5 Exercises

3.1 Let (X, F) be an fts invertible with respect to g, and let $supp\, g = A \in F$. Prove that A is also an inverting subset of (X, F).

3.2 Prove that every fts with a closed weak fuzzy point is invertible.

3.3 Prove that every quasi-separated fts is invertible.

3.4 Prove that every H-fts containing a well-closed fuzzy point is invertible.

3.5 Prove that every H-fts with a closed fuzzy point is invertible.

3.6 Let (X, F) be a T_1 H-fts containing a g-closed fuzzy point. Prove that (X, F) is invertible.

3.7 Prove that every FT_1 H-fts is invertible.

3.8 Let (X, F) be an H-fts containing two fuzzy points x_λ and x_γ such that $\lambda < \gamma$ and x_γ is open. Prove that (X, F) is invertible provided x_λ is g-closed.

3.9 Prove that any fts (X, F) such that $\underline{\alpha} \in F$ for some $\alpha \in [\frac{1}{2}, 1)$ is invertible.

3.10 Prove that every fully stratified fts is invertible.

3.11 Prove or disprove a purely stratified fts is invertible.

3.12 Prove that every topologically generated fts is invertible.

3.13 Let (X, F) be an fts and $f \in F$ be a weak fuzzy subset of F which contains a g-closed fuzzy subset. Prove that (X, F) is invertible.

3.14 Prove that any fts (X, F) containing an open fuzzy point, with $|X| > 2$ is not completely invertible.

3.15 Prove that a fully stratified fts is not completely invertible.

3.16 Prove that a homogeneous finite c-fts is always invertible.

3.17 Let (X, F) be an fts invertible with respect to a crisp subset A. If (A, F_A) is strongly homogeneous, then show that (X, F) is also strongly homogeneous.

3.18 Prove that a completely invertible finite c-fts (X, F) is strongly homogeneous.

3.19 Prove that in a completely invertible FT_1 fts, orbits are not finite.

3.20 Prove that in any fuzzy topological space, an orbit can never be an inverting subset.

References

1. Mathew, S.C.: Invertible fuzzy topological spaces. STARS: Int. J. **7**, 1–13 (2006)
2. Mathew, S.C., Jose, A.: Invertible and completely invertible fuzzy topological spaces. J. Fuzzy Math. **18**(3), 647–658 (2010)
3. Jose, A., Mathew, S.C.: Orbits and homogeneities of invertible fuzzy topological spaces. J. Adv. Stud. Topol. **3**(3), 23–31 (2012)

Chapter 4
Types of Invertible Fuzzy Topological Spaces

The chapter closely examines the structure of the inverting pairs and indicates the possible ones. It dissects invertible fuzzy topological space in an interesting manner based on inverting maps and their characterizing properties are derived. The conditions under which a homogeneous fuzzy topological space becomes type 1 invertible are also discussed. The characterization of a completely invertible finite c-fuzzy topological spaces divides completely invertible finite fuzzy topological spaces into two classes.

4.1 Inverting Pairs

Certain properties of inverting pairs—inverting fuzzy set and its inverting map—are discussed here.

Definition 4.1 [1] If an fts (X, F) is invertible, then there exists an inverting fuzzy subset g and an inverting map θ of (X, F). This g and θ together called an inverting pair of (X, F).

Clearly there can be different inverting pairs for an invertible fts.

Theorem 4.1 [1] *Let θ be a homeomorphism of an fts (X, F) and $g \in F$; $g \neq \underline{0}, \underline{1}$. Then the invertibility of (X, F) with respect to the following pairs are equivalent.*

- *(i)* (g, θ).
- *(ii)* (g, θ^{-1}).
- *(iii)* $(\theta^{-1}(g), \theta^{-1})$.
- *(iv)* $(\theta(g), \theta)$.
- *(v)* $(\theta(g), \theta^{-1})$.
- *(vi)* $(\theta^{-1}(g), \theta)$.

A. Jose and S. C. Mathew, *Invertible Fuzzy Topological Spaces*,
https://doi.org/10.1007/978-981-19-3689-0_4

Proof. The assertions of the theorem follow from the following set of inequalities.

$$\theta(\mathcal{C}(g)) \le g \Rightarrow \mathcal{C}(g) \le \theta(g) \Rightarrow \theta^{-1}(\mathcal{C}(g)) \le g$$
$$\Rightarrow \mathcal{C}(\theta^{-1}(g)) \le g \Rightarrow \theta^{-1}(\mathcal{C}(\theta^{-1}(g))) \le \theta^{-1}(g)$$
$$\Rightarrow \mathcal{C}(\theta^{-1}(g)) \le g \Rightarrow \mathcal{C}(g) \le \theta(g) \Rightarrow \theta(\mathcal{C}(\theta(g))) \le \theta(g)$$
$$\Rightarrow \mathcal{C}(\theta(g)) \le \theta(\theta(g)) \Rightarrow \theta^{-1}(\mathcal{C}(\theta(g))) \le \theta(g)$$
$$\Rightarrow \mathcal{C}(\theta(g)) \le g \Rightarrow \theta(\mathcal{C}(\theta^{-1}(g))) \le \theta^{-1}(g)$$
$$\Rightarrow \mathcal{C}(\theta(g)) \le g \Rightarrow \theta(\mathcal{C}(g)) \le g. \qquad \square$$

The following is an example of an invertible fts where the above 6 inverting pairs are different.

Example 4.1 Let $X = \{a, b, c, d\}$. Consider the fuzzy sets $f, g, h, i \in I^X$ defined by

$$f(a) = \tfrac{1}{3}, \quad f(b) = \tfrac{1}{4}, \quad f(c) = \tfrac{6}{7}, \quad f(d) = \tfrac{7}{8}$$
$$g(a) = \tfrac{6}{7}, \quad g(b) = \tfrac{7}{8}, \quad g(c) = \tfrac{1}{4}, \quad g(d) = \tfrac{1}{3}$$
$$h(a) = \tfrac{7}{8}, \quad h(b) = \tfrac{6}{7}, \quad h(c) = \tfrac{1}{3}, \quad h(d) = \tfrac{1}{4}$$
$$i(a) = \tfrac{1}{4}, \quad i(b) = \tfrac{1}{3}, \quad i(c) = \tfrac{7}{8}, \quad i(d) = \tfrac{6}{7}.$$

Let F be the fuzzy topology generated by $\{f, g, h, i\}$. Define $\theta : X \to X$ by $\theta(a) = c$, $\theta(b) = d$, $\theta(c) = b$, $\theta(d) = a$. Clearly (g, θ), (g, θ^{-1}), $(\theta^{-1}(g), \theta^{-1})$, $(\theta(g), \theta)$, $(\theta(g), \theta^{-1})$, $(\theta^{-1}(g), \theta)$ are different inverting pairs of (X, F).

Corollary 4.1 [1] *Let* (X, F) *be an invertible fts and* (g, θ) *be an inverting pair of* (X, F). *Then* $\mathcal{C}(g) \le \theta(g) \wedge \theta^{-1}(g)$

Proof. Since (g, θ) be an inverting pair of (X, F), $\mathcal{C}(g) \le \theta^{-1}(g)$. By above theorem (g, θ^{-1}) is also an inverting pair of (X, F) so that $\mathcal{C}(g) \le \theta(g)$. $\qquad \square$

If (g, e) is an inverting pair of an fts (X, F) then the following theorem gives a clear picture of the structure of g.

Theorem 4.2 [1] *Let* (X, F) *be an invertible fts. Then* (g, e) *is an inverting pair of* (X, F) *if and only if* $\tfrac{1}{2} \le g$.

Proof. Obvious. $\qquad \square$

Theorem 4.3 [1] *Let* (X, F) *be an invertible fts and* (g, θ) *be an inverting pair of* (X, F). *Then the following statements are equivalent.*

 (i) *g and* $\theta(g)$ *are not quasi-coincident.*
 (ii) $\theta(g) = \mathcal{C}(g)$.
 (iii) $\theta^{-1}(g) = \mathcal{C}(g)$.

Proof. (i) $\Longleftrightarrow$ (ii) Since θ is an inverting map for g, $\theta(\mathcal{C}(g)) \le g$ so that by theorem 1.4, $\theta(\mathcal{C}(g))$ and $\mathcal{C}(g)$ are not quasi-coincident. Then $\theta(\mathcal{C}(g))(x) + \mathcal{C}(g)(x) \le 1$ for all x in X so that $\theta(g)(x) + g(x) \ge 1$ for all x in X. Hence $\theta(g)(x) + g(x) = 1$ for all x in X if and only if $\theta(g)(x) + g(x) \not> 1$ for any x in X. Consequently $\theta(g) = \mathcal{C}(g)$ if and only if g and $\theta(g)$ are not quasi-coincident.

(ii) $\Longleftrightarrow$

(iii) $\theta(g) = \mathcal{C}(g)$ if and only if $g = \theta^{-1}(\mathcal{C}(g))$ if and only if $\mathcal{C}(g) = \theta^{-1}(g)$.

$\square$

Theorem 4.4 [1] *Let (X, F) be an invertible fts and (g, θ) be an inverting pair of (X, F). If g and $\theta(g)$ are not quasi-coincident then*

(i) $\theta(g) = \theta^{-1}(g)$.

(ii) g *is closed in* (X, F).

(iii) $\theta^2 = e$ *if $g(a) \neq g(b)$ for distinct $a, b \in X$.*

(iv) (X, F) *is a complete H-fts provided $g(x) \neq \frac{1}{2}$ for all $x \in X$.*

(v) $|X|$ *is even provided X is finite and $g(x) \neq \frac{1}{2}$ for all $x \in X$.*

Proof. (i) By theorem 4.3, g and $\theta(g)$ are not quasi-coincident if and only if $\theta(g) = \mathcal{C}(g) = \theta^{-1}(g)$.

(ii) Follows from theorem 4.3(ii).

(iii) From (i), we have $\theta(g) = \theta^{-1}(g)$. But we claim that $\theta = \theta^{-1}$. For, if $\theta \neq \theta^{-1}$, then there exists at least one $x \in X$ such that $\theta(x) \neq \theta^{-1}(x)$. Let $\theta(x) = y$ and $\theta^{-1}(x) = z$ where $y \neq z$. Then $g(z) = g(\theta^{-1}(x)) = \theta(g)(x) = \theta^{-1}(g)(x) = g(\theta(x)) = g(y)$, a contradiction. Hence the claim so that $\theta^2 = e$.

(iv) Since g and $\theta(g)$ are not quasi-coincident by theorem 4.3, $\theta(g) = \mathcal{C}(g)$. Since $g(x) \neq \frac{1}{2}$, $\forall x \in X$, this implies that $\theta(g)(x) \neq g(x)$ for all x in X. Consequently $\theta(x) \neq x$ for all x in X so that (X, F) is a complete H-fts.

(v) By theorem 4.3, $\theta(g) = \mathcal{C}(g)$ so that for any $x \in X$, $1 - g(x) = \theta(g)(x) = g(y)$ where $\theta(y) = x$. Clearly $y \neq x$. Thus for any $x \in X$, there exists a $y \in X$ with $y \neq x$ such that $g(x) + g(y) = 1$. Since $g(x) \neq \frac{1}{2}$, $\forall x \in X$, this implies that $|X|$ is even. $\square$

Remark 4.1 None of the reverse implications in the above theorem are true. For example, let $X = \{a, b, c, d\}$. Consider the fuzzy subsets f, g, h, i, j, k, l, n of X defined by

$$f(a) = \tfrac{2}{3}, \quad f(b) = \tfrac{3}{4}, \quad f(c) = \tfrac{4}{5}, \quad f(d) = \tfrac{5}{6}$$
$$g(a) = \tfrac{3}{4}, \quad g(b) = \tfrac{2}{3}, \quad g(c) = \tfrac{5}{6}, \quad g(d) = \tfrac{4}{5}$$
$$h(a) = \tfrac{2}{3}, \quad h(b) = \tfrac{2}{3}, \quad h(c) = \tfrac{4}{5}, \quad h(d) = \tfrac{4}{5}$$
$$i(a) = \tfrac{3}{4}, \quad i(b) = \tfrac{3}{4}, \quad i(c) = \tfrac{5}{6}, \quad i(d) = \tfrac{5}{6}$$
$$j(a) = \tfrac{1}{3}, \quad j(b) = \tfrac{1}{4}, \quad j(c) = \tfrac{1}{5}, \quad j(d) = \tfrac{1}{6}$$
$$k(a) = \tfrac{1}{4}, \quad k(b) = \tfrac{1}{3}, \quad k(c) = \tfrac{1}{6}, \quad k(d) = \tfrac{1}{5}$$
$$l(a) = \tfrac{1}{3}, \quad l(b) = \tfrac{1}{3}, \quad l(c) = \tfrac{1}{5}, \quad l(d) = \tfrac{1}{5}$$
$$n(a) = \tfrac{1}{4}, \quad n(b) = \tfrac{1}{4}, \quad n(c) = \tfrac{1}{6}, \quad n(d) = \tfrac{1}{6}.$$

Then (X, F) is an fts where $F = \{\underline{0}, \underline{1}, f, g, h, i, j, k, l, n\}$. Consider $\theta : X \to X$ defined by $\theta(a) = b$, $\theta(b) = a$, $\theta(c) = d$, $\theta(d) = c$ so that $\theta(g) = \theta^{-1}(g)$. Clearly

(g, θ) is an inverting pair of (X, F) and $\theta^2 = e$. Also g is closed in (X, F). Further, (X, F) is a complete H-fts. However, note that g and $\theta(g)$ are quasi-coincident.

Theorem 4.5 [1] *Let (X, F) be an invertible fts where X is finite and (g, θ) be an inverting pair such that g and $\theta(g)$ are not quasi-coincident. Then θ is the only inverting map for g if $g(a) \neq g(b)$ for distinct $a, b \in X$*

Proof. From theorem 4.3, $\theta(g) = \mathcal{C}(g)$. If possible let $\eta \neq \theta$ be an inverting map of g. Then by corollary 4.1, $\mathcal{C}(g) \leq \eta(g) \Rightarrow \theta(g) \leq \eta(g)$. We claim that $\theta(g) = \eta(g)$. Consider $S = \{x \in X : \theta(g)(x) < \eta(g)(x)\}$ and choose $y \in S$ such that $\theta(g)(y) = \min\{\theta(g)(x) : x \in S\}$. Since η is also a homeomorphism of (X, F), we have $\theta(g)(y) = \eta(g)(z)$ for some $z \in X$. But $y \neq z$ as $y \in S$. Then $\theta(g)(z) \not< \eta(g)(z)$. For, if $\theta(g)(z) < \eta(g)(z)$ then $z \in S$ and $\theta(g)(z) < \theta(g)(y)$, a contradiction. Also $\theta(g)(z) \neq \eta(g)(z)$. For, $\theta(g)(z) = \eta(g)(z) \Rightarrow \theta(g)(z) = \theta(g)(y)$, a contradiction since y and z are distinct. So we must have $\theta(g)(z) > \eta(g)(z)$ which is not possible. Consequently $S = \emptyset$ and hence the claim. Since $g(a) \neq g(b)$ for distinct $a, b \in X$, $\theta(g) = \eta(g) \Rightarrow \theta = \eta$. Thus θ is the only inverting map for g. $\square$

4.2 Type 1 and Type 2 Invertible Fuzzy Topological Spaces

Depending on the role of identity as an inverting map here we introduce two different types of invertible fuzzy topological spaces. Apart from characterizations of both the types we obtain several other properties enjoyed by these spaces. With regard to this division, a special attention is given to completely invertible fuzzy topological spaces.

Definition 4.2 [1] An invertible fts (X, F) is said to be type 1 if identity is an inverting map.

Definition 4.3 [1] An invertible fts (X, F) is said to be type 2 if identity is an inverting map for all the inverting fuzzy subsets.

Remark 4.2 Clearly every type 2 invertible fts is type 1 invertible, but the converse is not true.

Example 4.2 Let X be the set of all real numbers. Let $F = \{\underline{\alpha}; \ \alpha \in [\frac{1}{2}, 1] \cup \{0\}\}$. Then clearly (X, F) is a completely invertible fts of type 2.

Example 4.3 Every purely stratified invertible fts (X, F) is type 2 invertible. Also a fully stratified fts is type 1 invertible but need not be type 2 invertible. For example, let $X = \{1, 2, ..., 10\}$ and consider the fts (X, F) where $F = I^X$. Clearly (X, F) is fully stratified and completely homogeneous. Consider $f \in I^X$ defined by

$$f(x) = \begin{cases} \frac{5}{6}; & 1 \leq x \leq 9 \\ \frac{1}{6}; & x = 10 \end{cases}$$

Clearly f is an inverting fuzzy subset of (X, F), but identity is not an inverting map for f.

Remark 4.3 Apart from c-fts, there are invertible fts which are not type 1. For example, Let $X = \{a, b, c, d\}$. Consider the fuzzy sets $f, g, h \in I^X$ defined by

$$f(a) = 1, \quad f(b) = 0, \quad f(c) = \tfrac{3}{4}, \quad f(d) = 1$$
$$g(a) = 0, \quad g(b) = 1, \quad g(c) = 1, \quad g(d) = \tfrac{3}{4}$$
$$h(a) = 0, \quad h(b) = 0, \quad h(c) = \tfrac{3}{4}, \quad h(d) = \tfrac{3}{4}$$

Then (X, F) is an fts where $F = \{\underline{0}, \underline{1}, f, g, h\}$. Clearly $\theta : X \to X$ defined by $\theta(a) = b$, $\theta(b) = a$, $\theta(c) = d$, $\theta(d) = c$ is an inverting map for f. Also identity is not an inverting map for any of the inverting fuzzy subsets so that (X, F) is invertible but not type 1.

A simple characterization for a type 1 invertible fts is

Theorem 4.6 [1] *An fts (X, F) is type 1 invertible if and only if there exists an $f \in F$ and $f \neq \underline{1}$ such that $\frac{1}{2} \leq f$.*

Proof. Follows from theorem 4.2. $\qquad\square$

Theorem 4.7 [1] *If an fts (X, F) is homogeneous and $X_\alpha \neq \emptyset$ for some $\alpha \in [\frac{1}{2}, \beta]$ where $\beta < 1$ and $X_\alpha = \emptyset$ for each $\alpha \in [\beta, 1]$, then (X, F) is type 1 invertible.*

Proof. Let (X, F) be a homogeneous fts and $X_\alpha \neq \emptyset$ for some $\alpha \in [\frac{1}{2}, \beta] : \beta < 1$ and $X_\alpha = \emptyset$ for each $\alpha \in [\beta, 1]$. Then by exercise 2.11, $X_\alpha = X$ so that for each $x \in X$ there exists an open fuzzy subset g_x such that $g_x(x) = \alpha$. Now let $h = \bigvee_{f \in F - \{\underline{1}\}} f$.

Since $X_\alpha = \emptyset$ for each $\alpha \in [\beta, 1]$, we have $h = \underline{\alpha}$ for some $\alpha \in [\frac{1}{2}, \beta]$. Hence (h, e) is an inverting pair of (X, F). Consequently (X, F) is type 1 invertible. $\qquad\square$

Theorem 4.8 [1] *If a finite fts (X, F) is homogeneous and $X_\alpha \neq \emptyset$ for some $\alpha \in [\frac{1}{2}, 1)$, then (X, F) is type 1 invertible.*

Proof. Let (X, F) be a homogeneous fts and $X_\alpha \neq \emptyset$ for some $\alpha \in [\frac{1}{2}, 1)$. Then by exercise 2.11, $X_\alpha = X$ so that for each $x \in X$ there exists an open fuzzy subset g_x such that $g_x(x) = \alpha$. Now for each $x \in X$, choose one open fuzzy subset, say f_x such that $f_x(x) = \alpha$ and Let $f = \bigvee_{x \in X} f_x$. Clearly $f \neq \underline{1}$ and $f \geq \underline{\alpha}$. Thus (f, e) is an inverting pair of (X, F). $\qquad\square$

Theorem 4.9 [1] *Let (X, F) be an invertible fts which is not type 1. Then (X, F) has at least two inverting fuzzy subsets.*

Proof. Let (g, θ) be an inverting pair of (X, F). Then by theorem 4.1, $\theta(g)$ is an inverting fuzzy subset of (X, F). Also by theorem 4.2, there exists an $x \in X$ such that $g(x) < \frac{1}{2}$. Then by corollary 4.1, $\theta(g)(x) > \frac{1}{2}$ so that $g \neq \theta(g)$. $\qquad\square$

An elegant characterization of a type 2 completely invertible fuzzy topological space that can be easily obtained from theorem 4.2 is the following:

Theorem 4.10 [1] *An fts* (X, F) *is type 2 completely invertible if and only if* $\frac{1}{2} \leq g$ *for every* $g \in F$ *and* $g \neq \underline{0}$.

Theorem 4.11 [1] *Let* (X, F) *be a completely invertible fts where* X *is finite. Then* (X, F) *is a c-fts if it is not type 2.*

Proof. Suppose (X, F) is not type 2. Then to prove the theorem, we need only to show that $X_\alpha = \emptyset$ for every $\alpha \in (0, 1)$. But by theorem 3.13, $X_\alpha = \emptyset$ for every $\alpha \in (0, \frac{1}{2})$. So it remains to show that $X_\alpha = \emptyset$ for $\alpha \in [\frac{1}{2}, 1)$. If possible suppose there exists an $\alpha \in [\frac{1}{2}, 1)$ such that $X_\alpha \neq \emptyset$ which means there exists a $g \in F$ with $g(y) = \alpha$ for some y in X. Since (X, F) is completely invertible and not type 2, by theorem 4.10 and theorem 3.13, $X_\lambda \neq \emptyset$ for $\lambda = 0$ so that there exists an $f \in F$ and $f \neq \underline{0}$ such that $f(x) = 0$ for some $x \in X$.

Case 1: $f(y) = 1$. Let $h = f \wedge g$. Then $h(x) = 0$ and $h(y) = \alpha$. Let $h_1 = h \wedge \theta(h)$ where θ is an inverting map for h. Then clearly $|supp\ h_1| \leq |X| - 2$ and $h_1(y) \in (0, 1)$. Let θ_1 be an inverting map for h_1 and $h_2 = h_1 \wedge \theta_1(h_1)$. Then $|supp\ h_2| \leq |X| - 4$ and $h_2(y) \in (0, 1)$. Proceeding like this after a certain stage we get an $h_i \in F$ such that $|supp\ h_i| < |X| - \frac{|X|}{2}$ and $h_i(y) \in (0, 1)$. Thus h_i is a non-empty open subset in (X, F) with $|supp\ h_i| < \frac{|X|}{2}$. Since (X, F) is completely invertible, this contradicts theorem 3.4.

Case 2: $f(y) = 0$. Let η be an inverting map for f. Then $\eta(f)(y) = 1$. Now let $m = \eta(f) \wedge g$. Then $m(z) = 0$ for some z in X and $m(y) = \alpha$. Now proceeding as in case 1, we arrive at a step which contradicts theorem 3.4.

Case 3: $f(y) = \beta$ for some $\beta \in [\frac{1}{2}, 1)$. Consider $h = f \wedge g$. Then $h(x) = 0$ and $h(y) = \gamma = \min\{\alpha, \beta\}$ where $\gamma \in [\frac{1}{2}, 1)$. Again proceeding as in case 1 we arrive at a contradiction to theorem 3.4.

Consequently there exists no $\alpha \in [\frac{1}{2}, 1)$ such that $X_\alpha \neq \emptyset$. Hence the theorem. $\square$

Remark 4.4 By the above theorem, every completely invertible fts (X, F) where X is finite, is either type 2 or a c-fts. Since a c-fts cannot be completely invertible of type 2, this classification is mutually exclusive.

Theorem 4.12 [1] *Let* (X, F) *be a type 2 completely invertible fts where* $\frac{1}{2} \notin F$. *Then every proper open fuzzy subset of* X *is quasi-coincident with itself.*

Proof. Since (X, F) is type 2 completely invertible and $\frac{1}{2} \notin F$, by theorem 4.10, for every $f \in F$ and $f \neq \underline{0}$, there exists an $x \in X$ such that $f(x) > \frac{1}{2}$ so that f is quasi-coincident with itself. $\square$

4.3 Exercises

4.1 Let (X, F) be an fts invertible with respect to g such that g doesn't contain any crisp singleton. Show that identity is an inverting map of (X, F).

4.2 Let (X, F) be a completely invertible fts. Prove that (X, F) is regular, if g and $\theta_g(g)$ are not quasi-coincident for all $g \in F$ where θ_g is an inverting map for g.

4.3 Prove that a type 2 completely invertible fts is always a D-fts.

4.4 Verify whether the converse of exercise 4.3 is true.

4.5 Prove that a type 2 completely invertible fts is separable.

4.6 Give an example of separable space which is not type 2 completely invertible.

4.7 Let (X, F) be a completely invertible fts which is not type 1. Then show that $X_\alpha \neq \emptyset$ where $\alpha = 1$.

4.8 Prove that a completely invertible fts where X is finite, is either type 2 or static.

4.9 Prove that a type 2 completely invertible fts (X, F) where $X_{\frac{1}{2}} = \emptyset$ is always a static fts with power one.

4.10 Prove that a completely invertible fts (X, F) where X is finite and $X_{\frac{1}{2}} = \emptyset$ is always a static fts.

4.11 Give an example of a completely invertible fts which is not type 2.

4.12 Let (X, F) be a type 2 completely invertible fts. Prove that every fuzzy point is an accumulation point of any fuzzy subset f of X with $|supp\ f| > 1$.

4.13 Prove that the converse of theorem 4.12 is not true.

Reference

1. Mathew, S.C., Jose, A.: On the structure of invertible fuzzy topological spaces. Adv. Fuzzy Sets and Syst. **40**(1), 67–79 (2010)

Chapter 5
Properties of Invertible Fuzzy Topological Spaces

The chapter investigates and speculates mainly the effect of invertibility on certain fuzzy topological properties. In other words, this chapter illustrates the importance of invertibility in fuzzy topological space. Certain properties of a subspace determined by an invertible crisp open set can be carried over to the parent fuzzy topological space. For some properties, complete invertibility of the fuzzy topological space is required for such an extension. The fuzzy topological properties discussed here are separation axioms, countability axioms, compactness and fuzzy connectedness.

5.1 Separation and Invertibility

Here certain conditions for an invertible fuzzy topological spaces are examined to satisfy some separation axioms. Throughout this section A stands for a non-empty crisp open subset in a given fts (X, F).

Definition 5.1 [1] An fts (X, F) is called weakly quasi-separated if for any two fuzzy points x_λ and y_γ in X with $x_\lambda \in \mathcal{C}(y_\gamma)$, there exists an $f \in F$ such that $x_\lambda \in f \leq \mathcal{C}(y_\gamma)$ or there exists a $g \in F$ such that $y_\gamma \in g \leq \mathcal{C}(x_\lambda)$.

Definition 5.2 [2] A fuzzy topological space is said to be FT_0 if for every pair of distinct fuzzy points x_λ and y_γ, there exists an open fuzzy subset f such that $x_\lambda \leq f \leq \mathcal{C}(y_\gamma)$ or $y_\gamma \leq f \leq \mathcal{C}(x_\lambda)$.

Theorem 5.1 [3] *Let (X, F) be an fts invertible with respect to A. If (A, F_A) is weakly quasi-separated, then (X, F) is also weakly quasi-separated.*

Proof Let θ be an inverting homeomorphism for A. Then $\theta^{-1}(A) \geq \mathcal{C}(A)$. Let (A, F_A) be weakly quasi-separated. Then $(\theta^{-1}(A), F_{\theta^{-1}(A)})$ is also weakly quasi-separated. Let x_λ and y_γ be two fuzzy points in X with $x_\lambda \leq \mathcal{C}(y_\gamma)$. Then the following three cases arise:

A. Jose and S. C. Mathew, *Invertible Fuzzy Topological Spaces*,
https://doi.org/10.1007/978-981-19-3689-0_5

Case 1: x_λ and $y_\gamma \in A$. Then since (A, F_A) is weakly quasi-separated, there exists an $f \in F_A$ such that $x_\lambda \in f \leq \mathcal{C}(y_\gamma)$ or there exists a $g \in F_A$ such that $y_\gamma \in g \leq \mathcal{C}(x_\lambda)$. Since A is open, f and $g \in F$. Consequently (X, F) is weakly quasi-separated.

Case 2: x_λ and $y_\gamma \in \theta^{-1}(A)$. Since $(\theta^{-1}(A), F_{\theta^{-1}(A)})$ is weakly quasi-separated, this is similar to case 1.

Case 3: Exactly one of x_λ, $y_\gamma \in A$. Without loss of generality, let $x_\lambda \in A$ and $y_\gamma \notin A$. Then $x_\lambda \in A \leq \mathcal{C}(y_\gamma)$ so that (X, F) is weakly quasi-separated. $\square$

The following theorem can be proved by making some obvious amendments in the proof of Theorem 5.1.

Theorem 5.2 [3] *Let (X, F) be an fts invertible with respect to A. If (A, F_A) is quasi-separated, then (X, F) is also quasi-separated.*

Definition 5.3 [1] An fts (X, F) is called separated if for any two fuzzy points x_λ and y_γ in X with $x_\lambda \in \mathcal{C}(y_\gamma)$, there exist $f, g \in F$ such that $x_\lambda \in f$ and $y_\gamma \in g$ with $f \leq \mathcal{C}(g)$.

Theorem 5.3 [3] *Let (X, F) be a completely invertible fts. If (A, F_A) is separated, then (X, F) is also separated.*

Proof Let θ be an inverting homeomorphism for A. Since (A, F_A) is separated, $(\theta^{-1}(A), F_{\theta^{-1}(A)})$ is also separated. Let x_λ and y_γ be any two fuzzy points in (X, F) such that $x_\lambda \leq \mathcal{C}(y_\gamma)$. Then the following cases arise:

Case 1: Both x_λ, $y_\gamma \in A$ or $\theta^{-1}(A)$. Without loss of generality assume that x_λ, $y_\gamma \in A$. Since (A, F_A) is separated, there exist $f, g \in F_A$ such that $x_\lambda \in f$ and $y_\gamma \in g$ with $f \leq \mathcal{C}(g)$. Since A is open, we have $f, g \in F$.

Case 2: Exactly one of x_λ, $y_\gamma \in A$. Without loss of generality, let $x_\lambda \in A$ and $y_\gamma \notin A$. If $A \wedge \theta^{-1}(A) = \underline{0}$, there is nothing to prove. Otherwise, let $f = A \wedge \theta^{-1}(A)$ and let η be an inverting homeomorphism for f. Then $\eta(x_\lambda)$, $\eta(y_\gamma) \in f \leq A$, and $\eta(x_\lambda) \leq \mathcal{C}(\eta(y_\gamma))$. Then by case 1, there exist open fuzzy subsets g_1, $g_2 \in F$ such that $\eta(x_\lambda) \in g_1$ and $\eta(y_\gamma) \in g_2$ with $g_1 \leq \mathcal{C}(g_2)$. Thus there exist $\eta^{-1}(g_1)$, $\eta^{-1}(g_2) \in F$ such that $x_\lambda \in \eta^{-1}(g_1)$ and $y_\gamma \in \eta^{-1}(g_2)$ with $\eta^{-1}(g_1) \leq \mathcal{C}(\eta^{-1}(g_2))$. Consequently (X, F) is separated. $\square$

Definition 5.4 [2] An fts (X, F) is said to be FT_2 if for every pair of distinct fuzzy points x_λ and y_γ, there exist open fuzzy subsets f and g such that $x_\lambda \leq f \leq \mathcal{C}(y_\gamma)$, $y_\gamma \leq g \leq \mathcal{C}(x_\lambda)$ and $f \leq \mathcal{C}(g)$.

Corollary 5.1 [3] *Let (X, F) be a completely invertible fts. If (A, F_A) is separated, (X, F) is FT_2.*

Proof Since every separated fts is FT_2, the corollary follows from Theorem 5.3. $\square$

Corollary 5.2 [3] *Let (X, F) be a completely invertible fts. If (A, F_A) is weakly quasi-separated and regular, then (X, F) is separated.*

Proof Follows from Exercise 5.3 and Theorem 5.3. □

The following example shows that if (X, F) is not invertible with respect to A, then it is not weakly quasi-separated, even if (A, F_A) is separated.

Example 5.1 Let X be the set of all natural numbers and let A be a non-empty finite subset of X. For each $g \in I^A$, define the fuzzy subset h_g of X by

$$h_g(x) = \begin{cases} g(x); & x \in A \\ 0; & x \notin A \end{cases}$$

Let F be the fuzzy topology generated by the collection $\{h_g : g \in I^A\}$. Then (X, F) is not invertible with respect to A. Clearly (A, F_A) is separated, but (X, F) is not even weakly quasi-separated.

Remark 5.1 The separated property of both (X, F) and (A, F_A) need not imply the invertibility of (X, F) with respect to A.

Definition 5.5 [4] An fts (X, F) is called a fuzzy quasi T_0 space if for every $x \in X$, and $\lambda \neq \mu$, either $x_\lambda \notin \overline{x_\mu}$ or $x_\mu \notin \overline{x_\lambda}$ where $\lambda, \mu \in [0, 1]$ and $\lambda \neq \mu$.

Theorem 5.4 [5] *Let (X, F) be an fts invertible with respect to A. If (A, F_A) is fuzzy quasi-T_0, then (X, F) is also fuzzy quasi-T_0.*

Proof Let θ be an inverting homeomorphism for A. Then $\theta^{-1}(A) \geq \mathcal{C}(A)$. Let (A, F_A) be a fuzzy quasi-T_0 space. Then $(\theta^{-1}(A), F_{\theta^{-1}(A)})$ is also fuzzy quasi-T_0. Let x_λ and x_γ be two fuzzy points in X with $\lambda \neq \mu$. Then the following two cases arise:
Case 1: x_λ and $x_\gamma \in A$. Then since (A, F_A) is fuzzy quasi-T_0, either $x_\lambda \notin \overline{x_\mu}^A$ or $x_\mu \notin \overline{x_\lambda}^A$. But $\overline{x_\mu}^X \leq \overline{x_\mu}^A \vee \mathcal{C}(A)$ and $\overline{x_\lambda}^X \leq \overline{x_\lambda}^A \vee \mathcal{C}(A)$ so that either $x_\lambda \notin \overline{x_\mu}^X$ or $x_\mu \notin \overline{x_\lambda}^X$. Consequently (X, F) is fuzzy quasi-T_0.
Case 2: x_λ and $x_\gamma \in \theta^{-1}(A)$. Since $(\theta^{-1}(A), F_{\theta^{-1}(A)})$ is fuzzy quasi-T_0, this is similar to case 1. □

Remark 5.2 If (X, F) is not invertible with respect to A, then it need not be fuzzy quasi-T_0, even if (A, F_A) is fuzzy quasi-T_0. For example, consider the fts (X, F) and the subspace (A, F_A) in example 5.1. Clearly (A, F_A) is fuzzy quasi-T_0, but (X, F) is not. Also (X, F) is not invertible with respect to A.

Definition 5.6 [4] An fts (X, F) is called a fuzzy T_0 space if for any two fuzzy points x_λ and y_γ such that $x_\lambda \neq y_\gamma$, either $x_\lambda \notin \overline{y_\gamma}$ or $y_\gamma \notin \overline{x_\lambda}$. Obviously every fuzzy T_0 space is fuzzy quasi-T_0.

Theorem 5.5 [5] *Let (X, F) be an fts invertible with respect to A. If (A, F_A) is fuzzy T_0 and all the fuzzy points in A are g-closed in (X, F), then (X, F) is also fuzzy T_0.*

Proof Let θ be an inverting homeomorphism for A. Then $\theta^{-1}(A) \geq \mathcal{C}(A)$. Let (A, F_A) be fuzzy T_0. Then $(\theta^{-1}(A), F_{\theta^{-1}(A)})$ is also fuzzy T_0. Let x_λ and y_γ be two fuzzy points in X with $x_\lambda \neq y_\gamma$. Then the following three cases arise:
Case 1: x_λ and $y_\gamma \in A$. Then since (A, F_A) is fuzzy T_0, either $x_\lambda \notin \overline{y_\gamma}^A$ or $y_\gamma \notin \overline{x_\lambda}^A$ so that $x_\lambda \notin \overline{y_\gamma}^X$ or $y_\gamma \notin \overline{x_\lambda}^X$. Consequently (X, F) is fuzzy T_0.
Case 2: x_λ and $y_\gamma \in \theta^{-1}(A)$. Since $(\theta^{-1}(A), F_{\theta^{-1}(A)})$ is also fuzzy T_0, this is similar to case 1.
Case 3: Exactly one of x_λ, $y_\gamma \in A$. Without loss of generality, let $x_\lambda \in A$ and $y_\gamma \notin A$. Since x_λ is g-closed in (X, F), $\overline{x_\lambda}^X \leq A$ so that $y_\gamma \notin \overline{x_\lambda}^X$. Consequently (X, F) is fuzzy T_0.   □

Example 5.2 Let X be the set of all rational numbers and A be a finite subset of X. For each $g \in I^A$, define $f_g, h_g \in I^X$ by

$$f_g(x) = \begin{cases} g(x); & x \in A \\ 1; & x \notin A \end{cases}$$

$$h_g(x) = \begin{cases} g(x); & x \in A \\ 0; & x \notin A \end{cases}$$

Let F be the fuzzy topology generated by $\{f_g, h_g : g \in I^A\}$. Then $A \in F$ and (X, F) is not invertible with respect to A and all the fuzzy points in A are generalized closed in (X, F). Clearly (A, F_A) is fuzzy T_0, but (X, F) is not.

Theorem 5.6 [3] *Let (X, F) be an fts invertible with respect to A. If (A, F_A) is T_1, then (X, F) is also T_1.*

Proof Let θ be an inverting homeomorphism for A. Then $\theta^{-1}(A) \geq \mathcal{C}(A)$. Let (A, F_A) be T_1. Then $(\theta^{-1}(A), F_{\theta^{-1}(A)})$ is also T_1. Let x_λ and y_γ be two distinct fuzzy points in X. Then the following four cases arise:
Case 1: x_λ and $y_\gamma \in A$. Then since (A, F_A) is T_1, there exist an $f \in F_A$ such that $x_\lambda \in f$ and $y_\gamma \notin f$ and also there exists a $g \in F_A$ such that $y_\gamma \in g$ and $x_\lambda \notin g$. Since A is open, f and $g \in F$. Consequently (X, F) is T_1.
Case 2: x_λ and $y_\gamma \in \theta^{-1}(A)$. Since $(\theta^{-1}(A), F_{\theta^{-1}(A)})$ is also T_1, this is similar to case 1.
Case 3: $x_\lambda \in A, x_\lambda \notin \theta^{-1}(A)$ and $y_\gamma \in \theta^{-1}(A), y_\gamma \notin A$. Since A and $\theta^{-1}(A)$ is open in X, (X, F) is T_1.
Case 4: $x_\lambda \in \theta^{-1}(A), x_\lambda \notin A$ and $y_\gamma \in A, y_\gamma \notin \theta^{-1}(A)$. This is similar to case 3. □

The following example shows that the conclusion of the above theorem is false if invertibility of (X, F) with respect to A is omitted from the hypothesis.

Example 5.3 Let X be the set of all natural numbers and A be any finite subset of X. For each $g \in I^X$ with $g(x) = 0$ for some x in A, define $f_g, h_g \in I^X$ by

$$f_g(x) = \begin{cases} g(x); & x \in A \\ \frac{1}{2}; & x \in \mathcal{C}(A) \end{cases}$$

$$h_g(x) = \begin{cases} g(x); & x \in A \\ 0; & x \in \mathcal{C}(A) \end{cases}$$

Then (X, F) is an fts where F is generated by the collection $\{f_g, h_g : g \in I^X$ with $g(x) = 0$ for some x in $A\}$. Then (A, F_A) is a T_1 fts, but (X, F) is not T_1. Note that $A \in F$ and (X, F) is not invertible with respect to A.

Definition 5.7 [4] An fts (X, F) is said to be fuzzy T_2 if for any two fuzzy points x_λ and y_γ with $x \neq y$, there exist Q-neighbourhoods f and g, respectively, of x_λ and y_γ such that $f \wedge g = \underline{0}$.

Theorem 5.7 [5] *Let (X, F) be a completely invertible fts. If (A, F_A) is fuzzy T_2, then (X, F) is also fuzzy T_2.*

Proof Let θ be an inverting homeomorphism for A. Since (A, F_A) is fuzzy T_2, $(\theta^{-1}(A), F_{\theta^{-1}(A)})$ is also fuzzy T_2. Let x_λ and y_γ be any two fuzzy points in (X, F) such that $x \neq y$. Then the following cases arise:
Case 1: Both x_λ, $y_\gamma \in A$ or $\theta^{-1}(A)$. Without loss of generality assume that x_λ, $y_\gamma \in A$. Since (A, F_A) is fuzzy T_2, there exist Q-neighbourhoods f and g of x_λ and y_γ in A, respectively, such that $f \wedge g = \underline{0}$. Since A is open, we have f and g are, respectively, Q-neighbourhoods of x_λ and y_γ in X also.
Case 2: Exactly one of x_λ, $y_\gamma \in A$. Without loss of generality, let $x_\lambda \in A$ and $y_\gamma \notin A$. If $A \wedge \theta^{-1}(A) = \underline{0}$, there is nothing to prove. Otherwise, let $f = A \wedge \theta^{-1}(A)$ and let η be an inverting homeomorphism for f. Then $\eta(x_\lambda)$, $\eta(y_\gamma) \in f \leq A$, and $\eta^{-1}(x) \neq \eta^{-1}(y)$. Then by case 1, there exist Q neighbourhoods g_1, g_2 in X, respectively, of $\eta(x_\lambda)$ and $\eta(y_\gamma)$ such that $g_1 \wedge g_2 = \underline{0}$. Thus, there exist $\eta^{-1}(g_1)$, $\eta^{-1}(g_2)$ such that $\eta^{-1}(g_1)$ is a Q-neighbourhood of x_λ and $\eta^{-1}(g_2)$ is a Q-neighbourhood of y_γ in X with $\eta^{-1}(g_1) \wedge \eta^{-1}(g_2) = \underline{0}$.
Consequently (X, F) is fuzzy T_2. □

Remark 5.3 Here also we note that the invertibility of (X, F) with respect to A is essential in the hypothesis of the theorem. For example, let X be any set and let A be a proper subset of X. For each $\alpha \in [\frac{1}{2}, 1]$ and each $x \in A$ define $f_{\alpha,x} \in I^X$ by

$$f_{\alpha,x}(y) = \begin{cases} \alpha; & y = x \\ 0; & y \neq x \end{cases}$$

Let F be the fuzzy topology generated by the collection $\{f_{\alpha,x} : x \in A, \ \alpha \in [\frac{1}{2}, 1]\}$. Then clearly (A, F_A) is fuzzy T_2, but (X, F) is not. Note that (X, F) is not invertible with respect to A.

Theorem 5.8 [5] *Let (X, F) be a completely invertible fts. If (A, F_A) is regular, then (X, F) is also regular.*

Proof Let x_λ be any fuzzy point in X and h be a closed fuzzy subset of X such that $x_\lambda \in \mathcal{C}(h)$. Let θ be an inverting homeomorphism for A. Since (A, F_A) is regular $(\theta^{-1}(A), F_{\theta^{-1}(A)})$ is also regular. Then there arise the following three cases:
Case 1: $x_\lambda \in A$ and not in $\theta^{-1}(A)$. Let $h_1 = h \wedge A$ and $h_2 = h \wedge (\theta^{-1}(A) \wedge \mathcal{C}(A))$ so that $h = h_1 \vee h_2$. Then h_1 is closed in A and h_2 is closed in $\theta^{-1}(A) \wedge \mathcal{C}(A) = \mathcal{C}(A)$ and hence in X. Since A is regular and open there exist two open fuzzy subsets f_1 and f_2 in A and hence in X such that $x_\lambda \in f_1$ and $h_1 \in f_2$ with $f_1 \leq \mathcal{C}(f_2)$. Also we claim that there exist two open fuzzy subsets f_3 and f_4 in X such that $x_\lambda \in f_3$ and $h_2 \in f_4$ with $f_3 \leq \mathcal{C}(f_4)$. For, if $A_1 = A \wedge \theta^{-1}(A) = \emptyset$, then $A = f_3$ and $\theta^{-1}(A) = f_4$. If $A_1 \neq \emptyset$, let η be an inverting homeomorphism for A_1, then $\eta(x_\lambda)$ and $\eta(h_2)$ are in $A_1 \leq A$. Since A is regular, there exists two open fuzzy subsets g_1 and g_2 in A such that $\eta(x_\lambda) \in g_1$ and $\eta(h_2) \in g_2$ with $g_1 \leq \mathcal{C}(g_2)$. Take $f_3 = \eta^{-1}(g_1)$ and $f_4 = \eta^{-1}(g_2)$ and hence the claim. Consequently, there exist two open fuzzy subsets $f_5 = f_1 \wedge f_3$ and $f_6 = f_2 \vee f_4$ in X such that $x_\lambda \in f_5$ and $h \in f_6$ with $f_5 \leq \mathcal{C}(f_6)$.
Case 2: $x_\lambda \in \theta^{-1}(A)$ and not in A. The proof is similar to case 1.
Case 3: $x_\lambda \in A \wedge \theta^{-1}(A)$. We have $h = h \wedge (A \vee \theta^{-1}(A)) = (h \wedge A) \vee (h \wedge \theta^{-1}(A))$. Since $h \wedge A$ is closed in A and A is regular there exists two open fuzzy subsets f_1 and f_2 in A such that $x_\lambda \in f_1$ and $h \wedge A \in f_2$ with $f_1 \leq \mathcal{C}(f_2)$. Similarly since $h \wedge \theta^{-1}(A)$ is closed in $\theta^{-1}(A)$ and $\theta^{-1}(A)$ is regular, there exists two open fuzzy subsets f_3 and f_4 in $\theta^{-1}(A)$ such that $x_\lambda \in f_3$ and $h \wedge \theta^{-1}(A) \in f_4$ with $f_3 \leq \mathcal{C}(f_4)$. Consequently, there exists two open fuzzy subsets $g_1 = f_1 \wedge f_3$ and $g_2 = f_2 \vee f_4$ in X such that $x_\lambda \in g_1$ and $h \in g_2$ with $g_1 \leq \mathcal{C}(g_2)$.
Consequently (X, F) is regular. $\square$

Corollary 5.3 [3] *Let (X, F) be a completely invertible fts. If (A, F_A) is quasi-separated and normal, then (X, F) is regular.*

Proof Follows from Exercise 5.4 and Theorem 5.8. $\square$

Theorem 5.9 [5] *Let (X, F) be a completely invertible fts. If (A, F_A) is normal, then (X, F) is also normal.*

Proof Let h_1 and h_2 be two closed fuzzy subsets in (X, F) such that $h_1 \leq \mathcal{C}(h_2)$. Then there arise two cases:
Case 1: $A \wedge \mathcal{C}(h_1 \vee h_2) \neq \underline{0}$. Let θ be an inverting homeomorphism for $A \wedge \mathcal{C}(h_1 \vee h_2)$. Then $\theta(h_i) \leq A \wedge \mathcal{C}(h_1 \vee h_2)$ for $i = 1, 2$ and $\theta(h_1) \leq \mathcal{C}(\theta(h_2))$. Then since A is normal, there exist two open fuzzy subsets f_1 and f_2 in A and hence in X such that $\theta(h_1) \leq f_1$ and $\theta(h_2) \leq f_2$ with $f_1 \leq \mathcal{C}(f_2)$. Then $h_1 \leq \theta^{-1}(f_1)$ and $h_2 \leq \theta^{-1}(f_2)$ with $\theta^{-1}(f_1) \leq \mathcal{C}(\theta^{-1}(f_2))$.
Case 2: $A \wedge \mathcal{C}(h_1 \vee h_2) = \underline{0}$, then $A \leq h_1 \vee h_2$. Then $A = (A \wedge h_1) \vee (A \wedge h_2)$ and $A \wedge h_1$ is closed in A and so is $A \wedge h_2$. Since $A \wedge h_1 \leq \mathcal{C}(A \wedge h_2)$, there exist two open fuzzy subsets g_1 and g_2 in A and hence in X such that $A \wedge h_1 \leq g_1$ and $A \wedge h_2 \leq g_2$ with $g_1 \leq \mathcal{C}(g_2)$. Also $h_1 \wedge \mathcal{C}(A)$ and $h_2 \wedge \mathcal{C}(A)$ are closed in X. Let θ be an inverting homeomorphism for A. Then $\theta(h_1 \wedge \mathcal{C}(A)) \leq A$ and $\theta(h_2 \wedge \mathcal{C}(A)) \leq A$. Then $\theta(h_1 \wedge \mathcal{C}(A))$ and $\theta(h_2 \wedge \mathcal{C}(A))$ are closed in A with $\theta(h_1 \wedge \mathcal{C}(A)) \leq \mathcal{C}(\theta(h_2 \wedge \mathcal{C}(A)))$ so that there exist two open fuzzy subsets g_3 and g_4 in A such that $\theta(h_1 \wedge$

$\mathcal{C}(A)) \leq g_3$ and $\theta(h_2 \wedge \mathcal{C}(A)) \leq g_4$ with $g_3 \leq \mathcal{C}(g_4)$. Thus, there exist two open fuzzy subsets $\theta^{-1}(g_3)$ and $\theta^{-1}(g_4)$ in X such that $h_1 \wedge \mathcal{C}(A) \leq \theta^{-1}(g_3)$ and $h_2 \wedge \mathcal{C}(A) \leq \theta^{-1}(g_4)$ with $\theta^{-1}(g_3) \leq \mathcal{C}(\theta^{-1}(g_4))$. Then $h_1 = (h_1 \wedge A) \vee (h_1 \wedge \mathcal{C}(A)) \leq g_1 \vee \theta^{-1}(g_3)$ and $h_2 = (h_2 \wedge A) \vee (h_2 \wedge \mathcal{C}(A)) \leq g_2 \vee \theta^{-1}(g_4)$ so that $h_1 \leq g_1 \vee \theta^{-1}(g_3) \wedge \mathcal{C}(h_2)$ and $h_2 \leq g_2 \vee \theta^{-1}(g_4) \wedge \mathcal{C}(h_1)$. Clearly $g_1 \vee \theta^{-1}(g_3) \wedge \mathcal{C}(h_2)$ and $g_2 \vee \theta^{-1}(g_4) \wedge \mathcal{C}(h_2)$ are open in X with $g_1 \vee \theta^{-1}(g_3) \wedge \mathcal{C}(h_2) \leq \mathcal{C}(g_2 \vee \theta^{-1}(g_4) \wedge \mathcal{C}(h_1))$. Consequently (X, F) is normal. $\square$

The following example shows that Theorems 5.8 and 5.9 do not hold without the complete invertibility of (X, F).

Example 5.4 Let X be the set of all integers and let A be any finite subset of X. For each $\alpha \in I$, define

$$g_{\alpha, \frac{1}{3}}(x) = \begin{cases} \alpha; & x \in A \\ \frac{1}{3}; & \text{otherwise} \end{cases}$$

$$g_{\alpha, 0}(x) = \begin{cases} \alpha; & x \in A \\ 0; & \text{otherwise} \end{cases}$$

$$g_{\alpha, \frac{4}{5}}(x) = \begin{cases} \alpha; & x \in A \\ \frac{4}{5}; & \text{otherwise} \end{cases}$$

Then $g_{\alpha, 0}, \ g_{\alpha, \frac{1}{3}}, \ g_{\alpha, \frac{4}{5}} \in I^X, \ \forall \alpha \in I$. Let F be the fuzzy topology generated by the collection $\{g_{\alpha, 0}, \ g_{\alpha, \frac{1}{3}}, \ g_{\alpha, \frac{4}{5}} \in I^X : \ \alpha \in I\}$. Here (X, F) is invertible with respect to $g_{1, \frac{4}{5}}$, but is not completely invertible. Clearly (A, F_A) is regular and normal, but (X, F) is not so.

5.2 Countability and Invertibility

If the inverting set as a subspace satisfies the first or second axioms of countability, then it is found that the invertible fuzzy topological space also satisfies the respective axioms of countability. It is also shown that the same conclusion holds for Q-first axiom of countability and separability of an invertible fuzzy topological space.

Throughout this section also A stands for a non-empty crisp open subset in a given fts (X, F).

Definition 5.8 [4] An fts (X, F) is said to satisfy the second axiom of countability or is said to be a C_{II} space if F has a countable base.

Theorem 5.10 [3] *Let (X, F) be an fts invertible with respect to A. If (A, F_A) satisfies the second axiom of countability C_{II}, then (X, F) also satisfies C_{II}.*

Proof Let θ be an inverting homeomorphism for A and $\{h_i\}$ be a countable base for A. Then $\{h_i\} \cup \{\theta^{-1}(h_i)\}$ is a countable family of open fuzzy subsets of X. Let f be any open fuzzy subset of X containing a fuzzy point x_λ.

Case 1: $x_\lambda \in A$. Then $A \wedge f$ is a open in (A, F_A) and contains x_λ. Then by Theorem 1.3, there exists an $h_j \in \{h_i\}$ such that $x_\lambda \in h_j \le A \wedge f \le f$.

Case 2: $x_\lambda \in \mathcal{C}(A)$. Then $\theta(x_\lambda) \in A$ and $A \wedge \theta(f)$ is an open fuzzy subset of A containing $\theta(x_\lambda)$. Then there exists an $h_j \in \{h_i\}$ such that $\theta(x_\lambda) \in h_j \le A \wedge \theta(f) \le \theta(f)$ which implies that $x_\lambda \in \theta^{-1}(h_j) \le f$.

Thus by Theorem 1.3, $\{h_i\} \cup \theta^{-1}\{h_i\}$ is a countable base for (X, F) so that (X, F) is second countable.   □

The following example shows that the second countability of (A, F_A) need not be transferred to (X, F) if (X, F) is not invertible with respect to A.

Example 5.5 Let X be the set of all real numbers and let $A = [-1, 1]$. For each $x \in B = X - (-1, 1)$, let g_x be the fuzzy point x_λ where $\lambda = \frac{1}{|x|}$. Let F be the fuzzy topology generated by $A \cup \{g_x : x \in B\}$. Clearly (X, F) is not invertible with respect to A and is not C_{II}. But (A, F_A) satisfies the second axiom of countability.

Definition 5.9 [4] An fts (X, F) is said to satisfy the first axiom of countability or to be C_I if and only if every fuzzy point in (X, F) has a countable neighbourhood base.

Theorem 5.11 [3] *Let (X, F) be an fts invertible with respect to A. If (A, F_A) satisfies the first axiom of countability C_I, then (X, F) also satisfies C_I.*

Proof Let x_λ be any fuzzy point in X and let f be a neighbourhood of x_λ in X.

Case 1: $x_\lambda \in A$. Let $\{g_i\}$ be a countable neighbourhood base for x_λ in A. We have $A \wedge f$ is a neighbourhood of x_λ in A so that there exists a $g_k \in \{g_i\}$ such that $x_\lambda \in g_k \le A \wedge f \le f$. Thus, X has a countable neighbourhood base at x_λ, namely, $\{g_i\}$.

Case 2: $x_\lambda \in \mathcal{C}(A)$. Let θ be an inverting homeomorphism for A. Then $\theta(x_\lambda) \in A$. Let $\{g_i\}$ be a countable neighbourhood base for $\theta(x_\lambda)$ in A. We have $A \wedge \theta(f)$ is a neighbourhood of $\theta(x_\lambda)$ in A so that there exists a $g_k \in \{g_i\}$ such that $\theta(x_\lambda) \in g_k \le \theta(f) \wedge A \le \theta(f)$. Then $x_\lambda \in \theta^{-1}(g_k) \le f$. Thus $\{\theta^{-1}(g_i)\}$ is a countable neighbourhood base for x_λ in X.

Consequently, (X, F) is first countable.   □

Example 5.6 If (X, F) is not invertible with respect to A, then the first countability of (A, F_A) need not be transferred to (X, F). For example, let X be the set of all real numbers and define $F = \{g \in I^X : supp\, \mathcal{C}(g) \text{ is finite}\} \cup \{\underline{0}\}$. Then (X, F) is an fts. Let $B \ne \underline{0}$ be a finite crisp subset of X. Consider the simple extension $(X, F(B))$. Clearly $(X, F(B))$ is not invertible with respect to B and B as a subspace of $(X, F(B))$ satisfies the first axiom of countability. But $(X, F(B))$ is not C_I.

Definition 5.10 [4] An fts (X, F) is said to satisfy the Q-first axiom of countability or to be Q-C_1 if every fuzzy point in (X, F) has a countable Q-neighbourhood base.

Theorem 5.12 [5] *Let (X, F) be an fts invertible with respect to A which as a subspace satisfies the Q-first axiom of countability. Then (X, F) satisfies Q-C_1.*

Proof Let x_λ be any fuzzy point in X and let f be a Q-neighbourhood of x_λ in X.

Case 1: $x_\lambda \in A$. Let $\{h_i\}$ be a countable Q-neighbourhood base for x_λ in A. We have $A \wedge f$ is a Q-neighbourhood of x_λ in A so that there exists an $h_k \in \{h_i\}$ such that $h_k \leq A \wedge f \leq f$. Thus, X has a countable Q-neighbourhood base at x_λ, namely, $\{h_i\}$.

Case 2: $x_\lambda \in \mathcal{C}(A)$. Let θ be an inverting map for A. Then $\theta(x_\lambda) \in A$. Let $\{g_i\}$ be a countable Q-neighbourhood base for $\theta(x_\lambda)$ in A. We have $A \wedge \theta(f)$ is a Q-neighbourhood of $\theta(x_\lambda)$ in A so that there exists a $g_k \in \{g_i\}$ such that $g_k \leq \theta(f) \wedge A \leq \theta(f)$. Then $\theta^{-1}(g_k)$ is a Q-neighbourhood of x_λ and $\theta^{-1}(g_k) \leq f$. Thus $\{\theta^{-1}(g_i)\}$ is a countable Q-neighbourhood base for x_λ in X.

Consequently, (X, F) satisfies the Q-first axiom of countability. □

Remark 5.4 If (X, F) is not invertible with respect to A, then the Q-first axiom of countability of (A, F_A) is not transferred to (X, F) in general. For example, let $X = [-1, 1]$. Let $F = \{g \in I^X; \; supp\mathcal{C}(g)$ is countable$\}$. Then (X, F) is an fts. Let B be a countable subset of X. Consider the simple extension $(X, F(B))$. Clearly $(B, F(B)_B)$ satisfies the Q-first axiom of countability, but $(X, F(B))$ is not. Clearly $(X, F(B))$ is not invertible with respect to B.

Theorem 5.13 [6] *Let (X, F) be an fts invertible with respect to a crisp subset A. Then (A, F_A) is separable implies (X, F) is separable.*

Proof Let $\{p_i\}$, $i = 1, 2, ...$ be a countable sequence of fuzzy points in (A, F_A) such that for every $f \in F_A$ and $f \neq \underline{0}$, there exists a p_i such that $p_i \in f$. Let θ be an inverting homeomorphism for A. Then $\{\theta^{-1}(p_i)\}$ is a countable sequence of fuzzy points in A and hence in X. Let $\{q_i\}$ be the sequence whose terms are alternatively the terms of $\{p_i\}$ and $\{\theta^{-1}(p_i)\}$. Then $\{q_i\}$ is a countable sequence of fuzzy points in (X, F). Let g be a non-empty open fuzzy subset of X. Then the following two cases arise:

Case 1: $g \wedge A \neq \underline{0}$. Then $g \wedge A$ is a non-empty open fuzzy subset in A so that there exists a $p_j \in \{p_i\}$ such that $p_j \in g \wedge A \leq g$.

Case 2: $g \wedge A = \underline{0}$. Then $g \leq \mathcal{C}(A)$ and since θ is an inverting homeomorphism for A, $\theta(g) \leq A$ and open in A. Then there exists a $p_k \in \{p_i\}$ such that $p_k \in \theta(g)$ so that $\theta^{-1}(p_k) \in g$.

Consequently, $\{q_i\}$ is a countable sequence of fuzzy points in (X, F) such that for every $f \in F$ and $f \neq 0$, there exists a q_i such that $q_i \in f$ so that (X, F) is separable. □

Remark 5.5 If A is not an inverting subset for (X, F), then (A, F_A) can be separable without (X, F) being separable. For, let X be the set of all real numbers and A be any countable subset of X. Define $F = \{f \in I^X : f \wedge A = \underline{\alpha}; \; \alpha \in [\frac{1}{2}, 1] \cup \{0\}\}$. Then (X, F) is an invertible fts which is not separable. Note that here (A, F_A) is separable and A is not an inverting subset of (X, F).

5.3   Compactness and Invertibility

This section probes the role of invertibility on compactness of fuzzy topological spaces.

Definition 5.11 [7] A family of open fuzzy subsets $\mathcal{U} = \{U_i : i \in \Delta\}$ of an fts (X, F) is an open covering of a fuzzy subset g if $g \leq \bigvee_{i \in \Delta} U_i$. A subcovering $\mathcal{V}$ of an open covering $\mathcal{U}$ of g is a subfamily $\mathcal{V}$ of $\mathcal{U}$ which is again an open covering of g.

Definition 5.12 [8] A fuzzy subset f of a fuzzy topological space (X, F) is compact if each open covering of f has a finite subcovering. An fts (X, F) is said to be a compact if $\underline{1}$ is compact.

Theorem 5.14 [8] *Let (X, F) be a compact fuzzy topological space and A be a closed crisp subset of (X, F). Then A is compact.*

Definition 5.13 [8] If every closed fuzzy subset of X is compact, then (X, F) is called strongly compact.

Definition 5.14 [9] Let (X, F) be an fts and let $\alpha \in I$. A collection $\mathcal{G} \subset F$ will be called an α-shading (resp., α^*-shading) of X if, for each $x \in X$ there exists a $g \in \mathcal{G}$ with $g(x) > \alpha$ (resp. $g(x) \geq \alpha$). A subcollection $\mathcal{H}$ of an α-shading (resp. α^*-shading) $\mathcal{G}$ of X that is also an α-shading (resp., α^*-shading) of $\mathcal{G}$ is called an α-subshading (resp., α^*-subshading) of $\mathcal{G}$. If each α-shading (resp. α^*-shading) of X has a finite α-subshading (resp. α^*-subshading), then (X, F) is called α-compact (resp. α^*-compact).

Theorem 5.15 [9] *The closed crisp subset of an α-compact (resp., α^*-compact) fts (X, F) is α-compact (resp., α^*-compact) as a subspace of (X, F).*

Proof Let $\mathcal{U}$ be an α-shading of B, and let $\mathcal{V} = \{V \in F : V|_B \in \mathcal{U}\}$. Then $\mathcal{V} \cup \{\mathcal{C}(B)\}$ is an α-shading of X, so $\mathcal{V} \cup \{\mathcal{C}(B)\}$ has a finite α-subshading $\{V_1, V_2, ..., V_n, \mathcal{C}(B)\}$. But then $\{V_1|_B, V_2|_B, ..., V_n|_B\}$ is a finite α-subshading of $\mathcal{U}$. Similarly, it can be proved for α^*-compact space. $\qquad\qquad\square$

Theorem 5.16 [7] *Let (X, F) and (Y, J) be two fuzzy topological spaces. If (X, F) is compact and $\theta : X \to Y$ is continuous and onto. Then (Y, J) is compact.*

Proof Let $\mathcal{U}$ be an open cover for Y. Then $\theta^{-1}(\mathcal{U})$ is an open cover for X. Since X is compact, it has a finite subcover, say $\mathcal{V}$. Then $\theta(\mathcal{V})$ is a finite subcover of $\mathcal{U}$ so that (Y, J) is compact. $\qquad\qquad\square$

Theorem 5.17 [10] *The continuous image of an α-compact (resp., α^*-compact) fts is α-compact (resp., α^*-compact).*

Proof Let (X, F) be an α-compact (resp., α^*-compact) space and $\theta : X \to Y$ be a continuous onto function from (X, F) to (Y, J). Let $\mathcal{G}$ be an α-shading (resp., α-shading) for Y. Then $\theta^{-1}(\mathcal{G})$ is an α-shading (resp., α^*-shading) for X. Since X is α-compact (resp., α-compact), it has a finite α-subshading (resp., α^*-subshading), say $\mathcal{H}$. Then $\theta(\mathcal{H})$ is a finite α-subshading (resp., α^*-subshading) of $\mathcal{G}$ so that (Y, J) is α-compact (resp., α^*-compact). $\qquad\square$

Remark 5.6 A type 2 completely invertible fts need not be compact. Let X be the set of all rational numbers. Then (X, F) where $F = \{\underline{0}, \underline{\alpha} : \alpha \in [\frac{6}{7}, 1]\}$ is a type 2 completely invertible fts. The family $\{\underline{\alpha} : \alpha \in [\frac{9}{10}, 1)\}$ is an open cover of $\underline{1}$, but it has no finite subcover. Hence, $\underline{1}$ is not compact. But all other closed fuzzy subsets are compact as ensured by the following theorem.

Theorem 5.18 [10] *Let (X, F) be a type 2 completely invertible fts and h be a closed fuzzy subset other than $\underline{1}$ of (X, F). Then h is compact.*

Proof From Theorem 4.10, it follows that $h \leq \frac{1}{2}$. Let $\mathcal{U}$ be any open cover of h. Since $\underline{0}$ is always compact, we may assume that $h \neq \underline{0}$. Then again by Theorem 4.10, any $g \in \mathcal{U}$ and $g \neq \underline{0}$ is such that $g \geq \frac{1}{2}$ and hence $\{g\}$ covers h so that h is compact. $\qquad\square$

As an instant consequence of the above theorem, one can obtain a characterization for type 2 compact completely invertible fts:

Corollary 5.4 [10] *A type 2 completely invertible fts is compact if and only if it is strongly compact.*

Consequently, under type 2 complete invertibility, the distinction between compactness and strong compactness vanishes.

Example 5.7 A strongly compact fts need not be invertible. Let X be the set of all rational numbers. Then (X, F) where $F = \{\underline{0}, \underline{1}, \underline{\alpha} : \alpha \in [\frac{1}{6}, \frac{1}{4}]\}$ is an fts. Clearly (X, F) is strongly compact, but is not invertible.

The following is an easy consequence of Theorem 4.10.

Theorem 5.19 [10] *A type 2 completely invertible fts is α-compact (resp., α^*-compact) for each $\alpha \in [0, \frac{1}{2})$ (resp., $\alpha \in [0, \frac{1}{2}])$.*

Remark 5.7 Converse of the above theorem is not true. Consider the fts (X, F) in Example 5.7. Clearly (X, F) is α-compact (resp., α^*-compact) for each $\alpha \in [\frac{1}{5}, 1)$ (resp., $\alpha \in [\frac{1}{5}, 1])$, but (X, F) is not invertible.

The following theorems explore the local to global nature of compactness, strongly compactness and α-compactness in an invertible fts.

Theorem 5.20 [10] *An fts (X, F) containing an invertible crisp subset A such that $\overline{A}$ is compact.*

Proof Since A is an inverting subset, there exists a homeomorphism θ of (X, F) such that $\theta(\mathcal{C}(A)) \leq A$. Since $\theta(\mathcal{C}(A))$ is a closed crisp subset of $\overline{A}$, it is compact by Theorem 5.14. Consequently $\mathcal{C}(A)$ is compact. Since $X = \overline{A} \vee \mathcal{C}(A)$ it follows that (X, F) is compact.   □

Theorem 5.21 [10] *If an fts (X, F) contains an invertible nearly crisp fuzzy subset A such that $(\overline{A}, F_{\overline{A}})$ is strongly compact, then X is compact.*

Proof Let (A, θ) be an inverting pair of (X, F) so that $\theta(\mathcal{C}(A)) \leq A$. Since $\theta(\mathcal{C}(A))$ is a closed fuzzy subset of the strongly compact fts $(\overline{A}, F_{\overline{A}})$, it is compact. Consequently $\mathcal{C}(A)$ is compact. Since $X = \overline{A} \vee \mathcal{C}(A)$ it follows that (X, F) is compact.   □

Theorem 5.22 [10] *If an fts (X, F) contains an invertible crisp subset A such that $(\overline{A}, F_{\overline{A}})$ is α-compact (resp., α-compact), then X is α-compact (resp., α^*-compact).*

Proof Since A is an inverting subset, there exists a homeomorphism θ of (X, F) such that $\theta(\mathcal{C}(A)) \leq A$. Since $\theta(\mathcal{C}(A))$ is a closed crisp subset of $\overline{A}$, it is α-compact (resp., α^*-compact) by Theorem 5.15. Consequently $\mathcal{C}(A)$ is α-compact (resp., α^*-compact). Since $X = \overline{A} \vee \mathcal{C}(A)$ it follows that (X, F) is α-compact (resp., α^*-compact).   □

Remark 5.8 Converse of Theorems 5.20 and 5.22 are not true. For example, let X be the set of all real numbers and define $G = \{g \in I^X : g(x) = 1$ for some $x \in X$ and there exists an $\epsilon > 0$ such that $g(y) = 1$, for all $y \in (x - \epsilon, x + \epsilon) = K$ and $g(x) = 0, \forall x \in \mathcal{C}(K)\}$. Let F be the fuzzy topology generated by $G \cup \{\frac{1}{2}\}$. Let $Y = [0, 1]$ and $F_Y = H$ be the subspace fuzzy topology on Y. Clearly (Y, F_Y) is compact and α-compact (resp., α^*-compact) for all $\alpha \in [0, 1]$. Let $A \in I^Y$ is defined by

$$A(x) = \begin{cases} 1; & x \in [0, 4/5) \\ 0; & x \in [4/5, 1] \end{cases}$$

Then A is an inverting fuzzy subset of (Y, H), but (A, H_A) is not compact. Moreover, (A, H_A) is not α-compact(resp., α^*-compact) for any $\alpha \in [0, 1)$ (resp., $(0, 1]$).

The following example shows that invertibility cannot be dropped from Theorems 5.20 and 5.22.

Example 5.8 Let X be the set of all natural numbers. Let $P, Q \in I^X$ by

$$P(x) = \begin{cases} 1; & x \text{ is odd} \\ 0; & \text{otherwise} \end{cases}$$

$$Q(x) = \begin{cases} 1; & x \text{ is even} \\ 0; & \text{otherwise} \end{cases}$$

For each $\alpha \in (0, 1]$, let $g_\alpha \in I^X$, defined by

$$g_\alpha(x) = \begin{cases} 1; & x \text{ is odd} \\ \alpha; & \text{otherwise} \end{cases}$$

Let F be the fuzzy topology generated by the collection
$\{P, Q, \{g_\alpha : \alpha \in (0, 1]\}, \{x_\lambda : \lambda = 1 \text{ and } x \text{ is even}\}\}$. Here $\overline{P} = P$ and clearly
$(\overline{P}, F_{\overline{P}})$ is compact, α-compact, and α^*-compact for all $\alpha \in [0, 1]$, but (X, F) is
not compact. Also (X, F) is not α-compact (resp., α^*-compact) for any $\alpha \in [0, 1)$
(resp., $(0, 1]$).

5.4 Connectedness and Invertibility

The effect of invertibility on connectedness of a fuzzy topological space is scrutinized
here.

Definition 5.15 [11] An [9] fts (X, F) is said to be fuzzy connected if it has no
proper fuzzy subset that is both open and closed.

Theorem 5.23 [11] *An fts (X, F) is fuzzy connected if and only if it has no non-empty
open fuzzy subsets f and g such that $f + g = \underline{1}$.*

Proof Suppose f and g are two open fuzzy subsets such that $f + g = \underline{1}$ then f is a
proper fuzzy set in X which is both open and closed in X. So X is not fuzzy connected.
If X is not fuzzy connected then it has a proper fuzzy subset f which is both closed
and open. So $g = \underline{1} - f$ is a fuzzy open set such that $g \neq 0$ and $f + g = \underline{1}$. □

Definition 5.16 [11] Let (X, F) be an fts, then $A \subset X$ is said to be fuzzy connected
subset if (A, F_A) is a fuzzy connected space.

Theorem 5.24 [11] *If (X, F) is an fts and A is a fuzzy connected subset of X, and
f and g are non-empty open fuzzy subsets of X such that $f + g = \underline{1}$, then either
$f \wedge A = \underline{1}_A$ or $g \wedge A = \underline{1}_A$.*

Proof Suppose there exists $x_0, y_0 \in A$ such that $f(x_0) \neq 1$ and $f(y_0) \neq 1$. Then
$f + g = \underline{1}$ implies that $f \wedge A + g \wedge A = \underline{1}$, where $f \wedge A \neq \underline{0}$ and $g \wedge A \neq \underline{0}$. So
by Theorem 5.23, A is not a fuzzy connected space. □

Definition 5.17 [12] Let (X, F) be an fts and $g \in I^X$, g is called a fuzzy connected
component of (X, F), if g is a maximally connected fuzzy subset of (X, F), i.e. if
for $f \in I^X$, f is fuzzy connected and $f \geq g \implies f = g$.

Definition 5.18 [11] An fts (X, F) is fuzzy super connected if the closure of every
non-empty open fuzzy subset of X is $\underline{1}$.

Theorem 5.25 [11] *If (X, F) is a fuzzy super connected fts then it is fuzzy connected.*

For a type 2 completely invertible fts, fuzzy connectedness and fuzzy super connectedness are equivalent.

Theorem 5.26 [10] *A type 2 completely invertible fts is fuzzy super connected if and
only if it is fuzzy connected.*

Proof The necessary part follows from Theorem 5.25.

Sufficiency: Suppose (X, F) is fuzzy connected. Then (X, F) has no proper fuzzy subset that is both open and closed so that $\frac{1}{2} \notin F$. Since (X, F) is type 2 completely invertible, by Theorem 4.10, $\frac{1}{2} \leq f$ for every $f \in F$ and $f \neq \underline{0}$. Hence, for every $f \in F$ and $f \neq \underline{0}, \frac{1}{2}, \overline{f} = \underline{1}$. Thus $\frac{1}{2} \notin F \implies (X, F)$ is fuzzy super connected. $\qquad\square$

But, in a type 2 invertible fts fuzzy connectedness need not imply fuzzy super connectedness as shown in the following example.

Example 5.9 Let X be the set of real numbers and $F = \{\underline{1}, \underline{0}\} \cup G \cup H$ where $G = \{\underline{\alpha} \in I^X : \alpha \in [\frac{5}{6}, \frac{6}{7}]\}$ and $H = \{\underline{\beta} \in I^X : \beta \in [\frac{1}{9}, \frac{1}{8}]\}$. Then clearly (X, F) is type 2 invertible and fuzzy connected. Now consider $f = \frac{1}{8}$. Then $\overline{f} = \frac{1}{7}$ and (X, F) is not fuzzy super connected.

Theorem 5.27 [10] *Let (X, F) be an invertible fts and (f, θ) be an inverting pair of (X, F). If f and $\theta(f)$ are not quasi-coincident then (X, F) is not fuzzy connected.*

Proof By Theorem 4.4, f is a proper fuzzy subset that is both open and closed so that (X, F) is not connected. $\qquad\square$

Theorem 5.28 [10] *A fuzzy connected finite c-fts is not completely invertible.*

Proof Follows from Theorems 3.11 and 5.27. $\qquad\square$

Remark 5.9 Converse of the above theorem is not true. For example, let X be the set of all real numbers and $F = \{\underline{\alpha} : \alpha \in [0, 1]\}$. Then (X, F) is an fts which is not completely invertible, but since $\frac{1}{2} \in F$, (X, F) is not fuzzy connected.

Theorem 5.29 [10] *Let (X, F) be a fuzzy connected fts and $\theta : X \to Y$ be a continuous onto function from (X, F) to (Y, G). Then (Y, G) fuzzy connected.*

Proof If possible assume that (Y, G) is not fuzzy connected. Then there exists two non-empty open fuzzy subsets g and h of Y such that $g + h = \underline{1}$. Hence, $\theta^{-1}(g)$ and $\theta^{-1}(h)$ are two non-empty open fuzzy subsets of X such that $\theta^{-1}(g) + \theta^{-1}(h) = \underline{1}$ so that (X, F) is not fuzzy connected, a contradiction. $\qquad\square$

The following theorem derives two components in a completely invertible fts which has a connected open fuzzy subset.

Theorem 5.30 [10] *Let A be a fuzzy connected open subset of a completely invertible fts (X, F). If (X, F) is not fuzzy connected, then A and $\mathcal{C}(A)$ are the homeomorphic components of (X, F).*

Proof Let θ be an inverting homeomorphism for A. Then by Theorem 4.1, $\mathcal{C}(A) \leq \theta(A)$. Since (X, F) is not fuzzy connected, there exists $g, h \in F$ and $g, h \neq \underline{0}$ such that $g + h = \underline{1}$. Let $g' = g \wedge A$ and $h' = h \wedge A$. Then by Theorem 5.24, one of them say $g' = 1_A$, so that $A \leq g$. Since $g + h = \underline{1}$, by Theorem 4.3, $\theta(g) = \mathcal{C}(g) = h$. Now $A \leq g \implies \theta(A) \leq h$ and $h \leq \mathcal{C}(A)$ so that $\theta(A) \leq \mathcal{C}(A)$. Hence $\mathcal{C}(A) = \theta(A)$. Consequently A and $\mathcal{C}(A)$ are the components of (X, F) and they are homeomorphic. $\qquad\square$

Theorem 5.31 [10] *If an FT_1 fts (X, F), with a non-empty fuzzy connected open subset A, is completely invertible, then (X, F) is fuzzy connected.*

Proof If (X, F) is not fuzzy connected, then by Theorem 5.30, A and $\mathcal{C}(A)$ are the components of (X, F) and they are homeomorphic. Let $x \in A$, then $B = A \wedge \mathcal{C}(x)$ is open. Let θ be an inverting homeomorphism for B. Then $\theta(\mathcal{C}(B)) \leq B$ so that $\theta(\mathcal{C}(A)) \leq B$. But then $\theta(\mathcal{C}(A))$ is a component of (X, F) properly contained in A, a contradiction. $\qquad\square$

5.5 Exercises

5.1 Let (X, F) be an fts invertible with respect to A. Prove that if (A, F_A) is weakly quasi-separated, then (X, F) is FT_0.

5.2 Let (X, F) be an fts invertible with respect to A. If (A, F_A) is quasi-separated, then prove that (X, F) is FT_1.

5.3 Prove that a weakly quasi-separated and regular fts is separated.

5.4 Prove that quasi-separated and normal fts is regular.

5.5 Let (X, F) be an fts invertible with respect to A. If (A, F_A) is fuzzy T_0 and all the fuzzy points in A are g-closed in (X, F), then show that (X, F) is fuzzy quasi-T_0.

5.6 If (X, F) be an fts invertible with respect to A and if (A, F_A) is quasi-separated, then prove that (X, F) is also quasi-separated.

5.7 Let (X, F) be a completely invertible fts. If (A, F_A) is quasi-separated and normal, show that (X, F) is separated.

5.8 Let (X, F) be a completely invertible space. If (A, F_A) is $T_{1\frac{1}{2}}$, then prove that (X, F) is $T_{1\frac{1}{2}}$.

5.9 Let (X, F) be a completely invertible fuzzy topological space. If (A, F_A) is strongly regular then show that (X, F) is also strongly regular.

5.10 Let (X, F) be completely invertible space. If (A, F_A) is completely normal then show that (X, F) is also completely normal.

5.11 Show by examples that complete invertibility cannot be dropped from Exercises 5.8 5.9 and 5.10.

5.12 Using Theorem 5.18, prove Corollary 5.4.

5.13 Show that every open fuzzy subset in a completely invertible strongly compact fts contains a compact fuzzy subset.

5.14 If a completely invertible fts (X, F) has an open crisp subset A such that (A, F_A) is strongly compact and regular, prove that X is normal.

References

1. Tsiporkova, E., Baets, B.D.: τ-compactness in separated fuzzy topological spaces. Tatra Mt. Math. Publ. **12**, 99–112 (1997)
2. Ghanim, M.H., Kerre, E.E., Mashhour, A.S.: Separation axioms, subspaces and sums in fuzzy topology. J. Math. Anal. Appl. **102**, 189–202 (1984)
3. Mathew, S.C., Jose, A.: On invertible fuzzy topological spaces. J. Fuzzy Math. **18**(2), 423–433 (2010)
4. Pu, P.M., Liu, Y.M.: Fuzzy topology. 1. neighbourhood structure of a fuzzy point and moore-smith convergence. J. Math. Anal. Appl. **76**, 571–599 (1980)
5. Mathew, S.C., Jose, A.: Some properties of invertible fuzzy topological spaces. Far East J. Math. Sci. **40**(1), 67–79 (2010)
6. Mathew, S.C., Jose, A.: Invertible and completely invertible fuzzy topological spaces. J. Fuzzy Math. **18**(3), 647–658 (2010)
7. Chang, C.L.: Fuzzy topological spaces. J. Math. Anal. Appl. **24**, 182–190 (1968)
8. Tsiporkova, E., Kerre, E.: On separation axioms and compactness in fuzzy topological spaces. J. Egypt. Math. Soc. **4**, 27–39 (1996)
9. Gantner, R.C., Steinlage, T.E., Warren, R.H.: Compactness in fuzzy topological spaces. J. Math. Anal. Appl. **62**, 547–562 (1978)
10. Jose, A., Mathew, S.C.: On compactness and connectedness of invertible fuzzy topological spaces. Int. J. Pure Appl. Math. **79**(4), 535–546 (2012)
11. Fatteh, U.V., Bassan, D.S.: Fuzzy connectedness and its stronger forms. J. Math. Anal. Appl. **III**, 449–464 (1985)
12. Palaniappan, N.: Fuzzy Topology. Narosa Publishing House, New Delhi (2002)

Chapter 6
Invertibility of the Related Spaces

The chapter examines the behaviour of subspaces, simple extensions and quotient spaces of invertible fuzzy topological spaces. The situations under which these spaces hold invertibility are also portrayed. A thorough probing on the invertibility of the sums and products of a family of invertible fuzzy topological spaces has been carried out focusing on different types of invertible fuzzy topological spaces. The associated spaces of invertible fuzzy topological spaces are also studied and some significant results are explored.

6.1 Sums and Subspaces

Here the additive and hereditary nature of invertibility of a fuzzy topological space is examined.

Theorem 6.1 *Invertibility is an additive property.*

Proof Let (X_i, F_i), $i \in \Delta$ be a family of pairwise disjoint invertible fuzzy topological spaces. For each $i \in \Delta$, let g_i be an inverting subset for (X_i, F_i). Let (X, F) be the sum fts. Define $g \in I^X$ by $g(x) = g_i(x) : x \in X_i$. Clearly $g \in F$. Since (X_i, F_i) is invertible with respect to g_i, there exists a homeomorphism θ_i of (X_i, F_i) such that $\theta_i(\mathcal{C}(g_i)) \le g_i$. Now define $\theta : X \to X$ by $\theta(x) = \theta_i(x)$, $x \in X_i$. Then clearly θ is a homeomorphism of (X, F). Let x be any element in X. Then $x \in X_i$ for some $i \in \Delta$ so that $\theta(\mathcal{C}(g))(x) = \theta_i(\mathcal{C}(g_i))(x) \le g_i(x) = g(x)$. Since x is arbitrary, we get $\theta(\mathcal{C}(g)) \le g$. Consequently (X, F) is invertible with respect to g. □

Remark 6.1 The sum fts of a family of pairwise disjoint type 2 invertible fuzzy topological spaces is invertible but need not be type 2. For example, let $X_1 = \{a, b\}$, $X_2 = \{c, d\}$, $X_3 = \{e, f\}$. Then (X_1, F_1), (X_2, F_2) and (X_3, F_3) are fuzzy topologies where $F_i = \{\underline{0}, \underline{1}, \frac{2}{3}, \frac{1}{3}\}$ where $i = 1, 2, 3$. Clearly all these fuzzy topological spaces

A. Jose and S. C. Mathew, *Invertible Fuzzy Topological Spaces*,
https://doi.org/10.1007/978-981-19-3689-0_6

are type 2 invertible. Now consider the sum fts (X, F) and $k \in I^X$ defined by
$k(a) = \frac{2}{3}$, $k(b) = \frac{2}{3}$, $k(c) = \frac{2}{3}$, $k(d) = \frac{2}{3}$, $k(e) = \frac{1}{3}$, $k(f) = \frac{1}{3}$.
Clearly $k \in F$. Define $\theta : X \to X$ by
$\theta(a) = a$, $\theta(b) = b$, $\theta(c) = e$, $\theta(d) = f$, $\theta(e) = c$, $\theta(f) = d$.
Clearly θ is a homeomorphism of (X, F). Also $\theta\mathcal{C}(k) \leq k$ so that (k, θ) is an inverting pair of (X, F). Hence, (X, F) is type invertible, but not type 2.

Remark 6.2 Complete invertibility is not an additive property.

Example 6.1 Let $X_1=\{a, b\}$, $X_2 = \{c, d\}$, $X_3=\{e, f\}$. Then (I^{X_1}, F_1), (I^{X_2}, F_2) and (I^{X_3}, F_3) are fuzzy topological spaces where $F_i = \{\underline{0}, \underline{1}, \frac{3}{4}\}$ where $i = 1, 2, 3$. Clearly all these fuzzy topological spaces are type 2 completely invertible. Now consider the sum space (I^X, F) and $g, h \in I^X$ defined by

$$g(a) = \tfrac{3}{4}, \; g(b) = \tfrac{3}{4}, \; g(c) = 0, \; g(d) = 0, \; g(e) = 0, \; g(f) = 0$$
$$h(a) = 1, \; h(b) = 1, \; h(c) = 0, \; h(d) = 0, \; h(e) = 1, \; h(f) = 1.$$

Clearly $g, h \in F$. Define $\theta : X \to X$ by
$\theta(a) = a$, $\theta(b) = b$, $\theta(c) = e$, $\theta(d) = f$, $\theta(e) = c$, $\theta(f) = d$. Clearly θ is a homeomorphism of (I^X, F). Also $\theta(\mathcal{C}(h)) \leq h$ so that (h, θ) is an inverting pair of (I^X, F). Hence (I^X, F) is invertible, but not type 2. Also g is not an inverting fuzzy subset of (I^X, F). Here note that the sum space is invertible with respect to $\frac{3}{4}$.

Remark 6.3 Type 2 invertibility, type 1 invertibility and invertibility are not hereditary properties. For example, let X be the set of all rational numbers. Consider $f \in I^X$ defined by

$$f(x) = \begin{cases} 1; & x \text{ is an integer} \\ \frac{1}{2}; & \text{otherwise} \end{cases}$$

Choose β such that $0 < \beta < \frac{1}{2}$. Let F be the fuzzy topology generated by $\{f, \underline{\alpha} : \alpha \in (0, \beta)\}$. Then (X, F) is a type 2 invertible fuzzy topological space. Let Y be the set of all integers. Clearly the subspace (Y, F_Y) is not invertible.

The following theorem gives a subclass of invertible fuzzy topological spaces for which type 1 invertibility is a hereditary property.

Theorem 6.2 [1] *Let (X, F) be a type 1 invertible fts with $X_\lambda = \emptyset$ for $\lambda = 1$. Then every subspace of (X, F) is also type 1 invertible.*

Proof Let (X, F) be a type 1 invertible fts and Y be a non-empty subset of X. By Theorem 4.6, there exists an $f \in F$ and $f \neq \underline{1}$ such that $f(x) \geq \frac{1}{2}$ for all x in X. Clearly $g = f \wedge Y$ is a proper open fuzzy subset of the subspace (Y, F_Y) and $g(y) \geq \frac{1}{2}$ for all y in Y. Then identity is an inverting map for g so that (Y, F_Y) is type 1 invertible.   □

6.2 Simple Extensions

In this section, the response of different types of invertible fuzzy topological spaces to simple extensions is noted and provided some situations under which the simple extensions do retain their original type.

Remark 6.4 The simple extension of a completely invertible fts need not be invertible. For, let $X = \{p, q, r, s\}$. Consider the fuzzy sets $f, h \in I^X$ defined by

$$f(p) = 1, \ f(q) = 0, \ f(r) = 1, \ f(s) = 0$$
$$h(p) = 0, \ h(q) = 1, \ h(r) = 0, \ h(s) = 1.$$

Then (X, F) is an fts where $F = \{\underline{0}, \underline{1}, f, h\}$. Clearly $\theta : X \to X$ defined by $\theta(p) = q$, $\theta(q) = p$, $\theta(r) = s$, $\theta(s) = r$ is an inverting map for both f and g. Hence (X, F) is completely invertible. Consider the simple extension $(X, F(g))$ where $g \in I^X$ defined by $g(p) = \frac{1}{3}$, $g(q) = \frac{1}{4}$, $g(r) = \frac{1}{5}$, $g(s) = \frac{1}{6}$. Clearly there exists no homeomorphism of $(X, F(g))$ other than the identity. Hence $(X, F(g))$ is not invertible.

Theorem 6.3 [1] *The simple extension of a type 1 invertible fts is type 1 invertible.*

Proof Let (X, F) be a type 1 invertible fts and $(X, F(g))$ be the simple extension of F determined by g where $g \in I^X$ and $g \notin F$. Let (f, e) be an inverting pair of (X, F). Since $F \subset F(g)$, (f, e) is also an inverting pair of $(X, F(g))$ so that $(X, F(g))$ is type 1 invertible. □

Corollary 6.1 [1] *The simple extension of a type 2 invertible fts is type 1 invertible.*

Proof Since every type 2 invertible fts is type 1 invertible, the result follows from Theorem 6.3. □

Remark 6.5 The simple extension of a type 2 invertible fts need not be type 2 invertible. Let X be the set of all natural numbers. Consider the fuzzy set $f \in I^X$ defined by

$$f(x) = \begin{cases} \frac{7}{9}; & x \text{ is odd} \\ \frac{2}{9}; & x \text{ is even} \end{cases}$$

Let F be the fuzzy topology generated by $\{f, \underline{\alpha} \in I^X : \alpha \in [\frac{1}{2}, 1]\}$. Then clearly (X, F) is a type 2 invertible fts.
Now consider the simple extension $(X, F(g))$ where $g \in I^X$ defined by

$$g(x) = \begin{cases} \frac{2}{9}; & x \text{ is odd} \\ \frac{7}{9}; & x \text{ is even} \end{cases}$$

Then $\theta : X \to X$ defined by

$$\theta(x) = \begin{cases} x + 1; \ x \text{ is odd} \\ x - 1; \ x \text{ is even} \end{cases}$$

is an inverting map for f in $(X, F(g))$ so that $(X, F(g))$ is invertible but not type 2.

Remark 6.6 The simple extension of a type 2 completely invertible fts need not be completely invertible. In a completely invertible fts no fuzzy point can be open. Consequently for any fuzzy point x_λ in X, $(X, F(x_\lambda))$ is not completely invertible. But in the case of a type 1 completely invertible fts, the simple extension is at least type 1 invertible as we see in the following theorem.

Theorem 6.4 [1] *The simple extension of a type 1 completely invertible fts is type 1 invertible.*

Proof Follows from Theorem 6.3.   □

Theorem 6.5 [1] *Let (X, F) be a type 2 completely invertible fts and $g \in I^X$ such that $g \notin F$. Then $(X, F(g))$ is type 2 completely invertible if and only if $g \geq \frac{1}{2}$.*

Proof From Theorem 4.10, the result follows.   □

Theorem 6.6 [1] *Let (X, F) be a type 2 completely invertible fts and $g \in I^X$ such that $g \notin F$. Then g is an inverting fuzzy subset of $(X, F(g))$ if and only if $g \geq \frac{1}{2}$.*

Proof If $g \geq \frac{1}{2}$, then by Theorem 6.5, $(X, F(g))$ is also type 2 completely invertible. So we may assume that there exists an $x \in X$ such that $g(x) < \frac{1}{2}$. Then we claim that g is not an inverting fuzzy subset of $(X, F(g))$. If (g, θ) is an inverting pair of $(X, F(g))$, then there exists a $y \in X$ and $y \neq x$ such that $\theta(x) = y$ and $\theta(g)(y) < \frac{1}{2}$. Also by Corollary 4.1, it follows that $g(y) > \frac{1}{2}$ so that $h(y) > \frac{1}{2}$, $\forall h \in F(g)$ and $h \neq \underline{0}$. Consequently $\theta(g) \notin F(g)$, which contradicts the fact that θ is a homeomorphism of $(X, F(g))$. Hence, g is an inverting fuzzy subset of $(X, F(g))$ if and only if $g \geq \frac{1}{2}$.
□

We have already seen that the type 2 nature of an invertible fts need not be carried over to its simple extensions. But it will be retained with simple extensions in the case of a type 2 completely invertible fts as proved in the following theorem.

Theorem 6.7 [1] *The simple extension of a type 2 completely invertible fts is type 2 invertible.*

Proof Let $(X, F(g))$ be the simple extension of a type 2 completely invertible fts (X, F), determined by $g \in I^X$ such that $g \notin F$. By Theorem 6.4, $(X, F(g))$ is type 1 invertible. If possible, let k be an inverting fuzzy subset of $(X, F(g))$ for which identity is not an inverting map. Clearly then $k \notin F$ and by Theorem 4.2, $k(x) < \frac{1}{2}$ for some $x \in X$. We have $k = f' \vee (f \wedge g)$ for some $f, f' \in F$ and (k, e) is not an inverting pair of $(X, F(g))$. Then by Theorem 4.10, we must have $f' = \underline{0}$ so that $k = f \wedge g$ with $g(x) < \frac{1}{2}$. Let θ be an inverting map for k, and let $\theta(x) = y$. Then $\theta(k)(y) = k(x) < \frac{1}{2}$ so that by Corollary 4.1, we have $k(y) > \frac{1}{2}$. This implies that $g(y) > \frac{1}{2}$ so that $h(y) > \frac{1}{2}$, $\forall h \in F(g)$ and $h \neq \underline{0}$. Consequently $\theta(k) \notin F(g)$, which contradicts the fact that θ is a homeomorphism of $(X, F(g))$. $\square$

6.3 Associated Spaces

A related problem here is to investigate the invertibility of the associated spaces. From the following example, it follows that the associated topological space of a completely invertible fts need not be invertible.

Example 6.2 Let X be the set of all integers and A be a finite subset of X. Then for each $\alpha \in [\frac{1}{2}, \frac{9}{10})$, define $f_\alpha \in I^X$ by

$$f_\alpha(x) = \begin{cases} \frac{11}{12}; & x \in A \\ \alpha; & x \notin A \end{cases}.$$

Let F be the fuzzy topology generated by $\{f_\alpha : \alpha \in [\frac{1}{2}, \frac{9}{10})\}$. Then (X, F) is completely invertible. But the associated topological space $(X, i(F))$, where $i(F) = \{\emptyset, X, A\}$, is not invertible.

Conversely, the complete invertibility of the associated topological space of an fts need not imply the invertibility of the fts as illustrated in the following example:

Example 6.3 Let X be the set of all natural numbers. For each $\alpha \in (0, \frac{1}{4})$, define $g_\alpha \in I^X$ by

$$g_\alpha(x) = \begin{cases} \alpha; & x \text{ is even} \\ \frac{1}{3}; & x \text{ is odd} \end{cases}.$$

For each $\beta \in (\frac{1}{4}, \frac{1}{3})$, define $h_\beta \in I^X$ by

$$h_\beta(x) = \begin{cases} \frac{4}{9}; & x \text{ is even} \\ \beta; & x \text{ is odd} \end{cases}.$$

Now consider the fuzzy topology generated by $\{g_\alpha : \alpha \in (0, \frac{1}{4})\} \cup \{h_\beta : \beta \in (\frac{1}{4}, \frac{1}{3})\}$. Then by Theorem 3.5, (X, F) is not invertible with respect to any $f \in F$. Let A and B denote the set of all odd and even integers, respectively. Then clearly $(X, i(F))$ where $i(F) = \{\emptyset, X, A, B\}$ is completely invertible.

Theorem 6.8 [2] *Any topologically generated fts is type 1 invertible but not completely invertible.*

Proof Every topologically generated fts is fully stratified. But a fully stratified fts is always type 1 invertible and is not completely invertible.   □

Remark 6.7 From Theorem 6.8, it follows that the associated fts of a given topological space (X, τ) is always invertible, irrespective of the invertibility of (X, τ). But the following theorem shows that the associated fts is type 2 invertible if and only if (X, τ) is not invertible.

Theorem 6.9 [2] *The associated fts (X, F) of a topological space (X, τ) is type 2 invertible if and only if (X, τ) is not invertible.*

Proof Suppose (X, F) is type 2 invertible. If possible, assume that (X, τ) is invertible. Let (A, θ) be an inverting pair of (X, τ), then by Theorem 1.12, θ is a homeomorphism of (X, F). Let $\alpha \in (\frac{1}{2}, 1)$ and consider $g \in I^X$ defined by

$$g(x) = \begin{cases} \alpha; & x \in A \\ 1 - \alpha; & \text{otherwise} \end{cases} .$$

Then clearly (g, θ) is an inverting pair of (X, F). Also here e cannot be an inverting map for g, so that (X, F) is not type 2 invertible, a contradiction.
Conversely suppose (X, τ) is not invertible. Since (X, F) is topologically generated, it is type 1 invertible. If possible let f be an inverting fuzzy subset of (X, F) for which e is not an inverting map. Then by Theorem 4.2, $f(y) < \frac{1}{2}$ for some $y \in X$. Now by Theorem 1.12, any inverting map θ of f is also a homeomorphism of (X, τ). Let $f(y) = \alpha$ and let $A = f^{-1}(\alpha, 1]$, then $A \neq \emptyset, \subset X$. Clearly (A, θ) is an inverting pair of (X, τ), a contradiction. Consequently (X, F) is type 2 invertible.   □

Remark 6.8 From the above proof, it follows that every inverting map of a topological space (X, τ) is always an inverting map of the associated fts (X, F). But conversely an inverting map of (X, F) need not be an inverting map of (X, τ). e is an inverting map for a type 1 invertible fts, but it cannot be an inverting map for the associated topological space. But from the converse part of the proof of the above theorem it is clear that an inverting map other than identity of an fts (X, F) is also an inverting map of the associated topological space.

Theorem 6.10 [2] *If the associated topological space $(X, i(F))$, where X is finite, is completely invertible then for each $g \in F$, there exists some $\alpha \in I$ such that $g \leq \underline{\alpha}$ with $|S_\alpha(g)| \geq \frac{|X|}{2}$.*

Proof Suppose the associated topological space $(X, i(F))$ of an fts (X, F) is completely invertible. Let g be an open fuzzy subset of X. If possible assume that for any $\alpha \in I$, $|S_\alpha(g)| < \frac{|X|}{2}$. Let $\lambda = \max_{x \in X} g(x)$ and $\beta = \max_{x \notin S_\lambda(g)} g(x)$. Now consider $A = f^{-1}(\beta, 1]$. Clearly $A = S_\lambda(g) \neq \emptyset$ and $A \in i(F)$. Then by Theorem 3.4, $(X, i(F))$ is not invertible with respect to A, a contradiction. Thus for any $g \in F$ there is an $\alpha \in I$ with $|S_\alpha(g) \geq \frac{|X|}{2}$. Choose α_0 be the maximum of all such $\alpha's$. Now if possible assume that $g \nleq \underline{\alpha_0}$. Then there is a $\gamma > \alpha$ such that $\gamma = g(x)$ for some

$x \in X$. Now consider $g^{-1}(\alpha_0, 1] = B$. Clearly $B \in i(F)$; $B \neq \emptyset$ and $(X, i(F))$ is not invertible with respect to B, a contradiction. $\qquad\qquad\qquad\qquad\qquad\square$

Remark 6.9 Converse of Theorem 6.10 is not true. Let $X = \{a, b, c, d\}$. Consider $g \in I^X$ defined by $g(a) = \frac{1}{2}$, $g(b) = \frac{2}{3}$, $g(c) = \frac{2}{3}$, $g(d) = \frac{2}{3}$. Then (X, F) is an fts where $F = \{\underline{0}, \underline{1}, g\}$. Here $g \leq \frac{2}{3}$ and $S_{\frac{2}{3}}(g) \geq \frac{|X|}{2}$. Clearly $(X, i(F))$ where $i(F) = \{X, \emptyset, \{b, c, d\}\}$ is not invertible.

6.4   Quotient Spaces

In general, the invertibility of a quotient space does not depend on the invertibility of the fuzzy topological space and vice versa. However, it is proved that the quotient space of a topologically generated fuzzy topological space is always type 1 invertible. But the quotient space of a completely invertible fuzzy topological space need not be invertible. Some conditions under which the quotient space of a completely invertible fuzzy topological space is completely invertible are also obtained.

Definition 6.1 [3] Let (X, F) be an fts and R be an equivalence relation on X. Let X/R be the quotient set and let $p : X \rightarrow X/R$ be the quotient map. Let G be the family of fuzzy subsets in X/R defined by $G = \{g : p^{-1}(g) \in F\}$. Then G is called the quotient fuzzy topology for X/R and $(X/R, G)$ is called the quotient fts.

First we note that type 2 complete invertibility is transferred to non-trivial quotient spaces.

Theorem 6.11 [2] *If an fts (X, F) is type 2 completely invertible then any non-trivial quotient space of (X, F) is type 2 completely invertible.*

Proof Let (X, F) be a type 2 completely invertible fts. Then by Theorem 4.10, $\frac{1}{2} \leq f$ for every $f \in F$ and $f \neq \underline{0}, \underline{1}$. Consider any quotient space $(X/R, G)$ of (X, F). Then for any $g \in G$ and $g \neq \underline{0}$, $g \geq \frac{1}{2}$ so that $(X/R, G)$ is type 2 completely invertible again by Theorem 4.10. $\qquad\qquad\qquad\square$

Remark 6.10 Converse of the above theorem is not true, in general. For example, let $X = [2, 3]$ and consider $f \in I^X$ defined by $f(x) = \frac{1}{x}$, $\forall x$. Then (X, F) is an fts where $F = \{\underline{0}, \underline{1}, f, \frac{2}{3}\}$. Clearly f is not an inverting fuzzy subset of (X, F). Let $(X/R, G)$ be any non-trivial quotient space of (X, F). Then clearly $G = \{\underline{0}, \underline{1}, \frac{2}{3}\}$ so that $(X/R, G)$ is type 2 completely invertible.

Remark 6.11 The quotient space of a completely invertible fts need not be invertible. Let X be the set of all natural numbers and let $Y = \{2x : x \in X\}$. For each $m \in Y$, define $f_m, g_m \in I^X$ by

$$f_m(x) = \begin{cases} \frac{1}{3}; & 1 \le x \le m \\ \frac{2}{3}; & \text{otherwise} \end{cases}$$

$$g_m(x) = \begin{cases} \frac{1}{3}; & m < x \le 2m \\ \frac{2}{3}; & \text{otherwise} \end{cases}$$

Let F be the fuzzy topology on X generated by $\{f_m, \; g_m : \; m \in Y\}$. Now consider for each $m \in Y$, $\theta_m : X \to X$ defined by

$$\theta_m(x) = \begin{cases} x + m; & 1 \le x \le m \\ x - m; & m < x \le 2m \\ x; & \text{otherwise.} \end{cases}$$

Clearly θ_m is a homeomorphism of (X, F). Also for each $m \in Y$, (f_m, θ_m) and (g_m, θ_m) are inverting pairs of (X, F) so that (X, F) is completely invertible. Now consider the equivalence relation R on X such that $X/R = \{A, B, C\}$ where $A = \{1\}$, $B = \{2\}$ and $C = \{x \in X : x > 2\}$. Consider $f \in I^{X/R}$ defined by $f(A) = \frac{1}{3}$, $f(B) = \frac{1}{3}$, $f(C) = \frac{2}{3}$. Then the quotient space $(X/R, G)$ of (X, F) is given by $G = \{\underline{0}, \underline{1}, f\}$ and is not invertible.

The following example shows that the quotient space of an fts may be completely invertible, even if the fts is not invertible.

Example 6.4 Let X be the set of all non-zero real numbers. For each $\alpha \in (\frac{2}{3}, 1]$, define $f_\alpha, g_\alpha \in I^X$ by

$$f_{\alpha(x)} = \begin{cases} \alpha; & x > 0 \\ 1 - \alpha; & \text{otherwise} \end{cases}$$

$$g_{\alpha(x)} = \begin{cases} 1 - \alpha; & x > 0 \\ \alpha; & \text{otherwise.} \end{cases}$$

Also consider $h \in I^X$ defined by

$$h(x) = \begin{cases} x; & 0 < x < 1 \\ 0; & -1 < x < 0, \\ \frac{1}{|x|}; & \text{otherwise.} \end{cases}$$

Let F be the fuzzy topology on X generated by $\{h, \; f_\alpha, \; g_\alpha \in I^X, \; \alpha \in (\frac{2}{3}, 1]\}$. Clearly the fts (X, F) is not invertible. Consider the equivalence relation R on X defined by $x R y$ if and only if $xy > 0$. Then the quotient space $(X/R, G)$ is completely invertible.

Now we identify certain fuzzy topological spaces whose quotient spaces are type 1 invertible.

Theorem 6.12 [2] *An fts with a type 1 invertible quotient space is type 1 invertible.*

Proof Let $(X/R, G)$ be a quotient fts of (X, F) and suppose it is type 1 invertible. Let (g, e) be an inverting pair of $(X/R, G)$. Then by Theorem 4.2, $\frac{1}{2} \leq g$. Now consider $f \in I^X$ defined by $f = p^{-1}(g)$ where p is the quotient map. Then clearly $f \in F$ such that $\frac{1}{2} \leq f$ and (f, e) is an inverting pair of (X, F). Hence, (X, F) is type 1 invertible. $\qquad \square$

Remark 6.12 There are type 1 invertible fts for which no non-trivial quotient space is invertible. Let X be the set of all natural numbers. Consider $f, g, h \in I^X$ defined by $f(x) = \frac{x}{x+1}$, $\forall x \in X$,

$$g(x) = \begin{cases} \frac{1}{3}; & x \text{ is even} \\ \frac{1}{5}; & x \text{ is odd} \end{cases}$$

$$h(x) = \begin{cases} \frac{1}{5}; & x \text{ is even} \\ \frac{1}{3}; & x \text{ is odd.} \end{cases}$$

Then (X, F) is an fts where $F = \{\underline{0}, \underline{1}, f, g, h, \frac{1}{3}, \frac{1}{5}\}$. Clearly (f, e) is an inverting pair of (X, F) so that (X, F) is type 1 invertible. Let $(X/R, G)$ be any non-trivial quotient space of (X, F). Clearly $m < \frac{1}{2}$, $\forall m \in G$ so that $(X/R, G)$ is not invertible.

Theorem 6.13 [2] *An fts (X, F) is invertible with $\underline{\alpha}$; $\alpha \in [\frac{1}{2}, 1)$ as an inverting fuzzy subset if and only if every quotient space of (X, F) is type 1 invertible.*

Proof Let $(X/R, G)$ be any quotient space of (X, F). Consider the fuzzy subset g of X/R defined by $g(D) = \alpha$, $\forall D \in X/R$. Clearly $g \in G$ and (g, e) is an inverting pair of $(X/R, G)$.
Conversely suppose every quotient space of (X, F) is type 1 invertible. Let R be an equivalence relation on X such that $X/R = \{X\}$. Consider the quotient space $(X/R, G)$. Since it is type 1 invertible, there exists an inverting fuzzy subset $f \in G$ such that $f(X) = \alpha$ for some $\alpha \in [\frac{1}{2}, 1)$. Let $g = p^{-1}(f)$, where p is the quotient map, then $g = \underline{\alpha}$ and $g \in F$ so that (g, e) is an inverting pair of (X, F). $\qquad \square$

6.5 Product Spaces

Finally, the product of a family of invertible fuzzy topological spaces is investigated and it is proved that the product space is invertible if at least one of the coordinate spaces is invertible. Further, the product space is type 2 completely invertible if and only if each coordinate space is type 2 completely invertible.

Definition 6.2 [3] Let (X_i, F_i), $i \in J$ be a family of fts'. Let X be the Cartesian product of $\{X_i, i \in J\}$ and let P_i be the projection of the product X into the ith coordinate set X_i. An $x \in X$ is of the form $(x_i, i \in J)$ where x_i is the ith component of x. Let $Q(J)$ denote the family of all finite subsets of J. Putting $B = \{\bigcap_{i \in K} P_i^{-1}(U_i) :$

$U_i \in F_i, K \in Q(J)\}$, we call the fuzzy topology F which takes B as a base, the product fuzzy topology for X, and B the defining base for the product fuzzy topology. The pair (X, F) is called the product space of the fts (X_i, F_i), $i \in J$.

Theorem 6.14 [2] *The product fuzzy topology (X, F) of a family of fts (X_i, F_i), $i \in J$ is invertible if (X_j, F_j) is invertible for some $j \in J$.*

Proof Let P_i be the projection of the product X into the ith coordinate set X_i and $\mathcal{B}$ be the defining base for (X, F). Suppose (X_j, F_j) is invertible for some $j \in J$. Let (g_j, θ_j) be an inverting pair of (X_j, F_j). Now consider $f = P_j^{-1}(g_j)$, clearly $f \in \mathcal{B}$ so that $f \in F$. Define $\theta : X \to X$ by $\theta(x) = y$ where $y = (y_i)$ such that $y_i = x_i$, $i \neq j$, $y_j = \theta_j(x_j)$. Clearly θ is a homeomorphism of (X, F). Let $x \in X$ and $\theta^{-1}(x) = z$. Then $\theta(\mathcal{C}(f))(x) = \mathcal{C}(f)(z) = \mathcal{C}(P_j^{-1}(g_j))(z) = \mathcal{C}(g_j)(z_j) = \mathcal{C}(g_j)(\theta_j^{-1}(x_j)) = \theta_j(\mathcal{C}(g_j))(x_j) \leq g_j(x_j) = P_j^{-1}(g)(x) = f(x)$ so that (f, θ) is an inverting pair of (X, F). $\qquad \square$

Theorem 6.15 [2] *The product fuzzy topology (X, F) of a family of fts' (X_i, F_i), $i \in J$ is type 1 invertible if (X_j, F_j) is type 1 invertible for some $j \in J$.*

Proof Let P_i be the projection of the product X into the ith coordinate set X_i and $\mathcal{B}$ be the defining base for (X, F). Suppose (X_j, F_j) is type 1 invertible for some $j \in J$. Let (g_j, e) be an inverting pair of (X_j, F_j). Then by Theorem 4.2, $g_j \geq \frac{1}{2}$. Now consider $P_j^{-1}(g_j) = f$, clearly $f \in F$ and $f \geq \frac{1}{2}$ so that e is an inverting map for f. Hence (X, F) is type 1 invertible. $\qquad \square$

Remark 6.13 The following example shows that the converse of Theorems 6.14 and 6.15 are not true. For each $\alpha \in [0, \frac{1}{2})$, let $X_\alpha = \{a, b\}$ and $f_\alpha(a) = \frac{1}{2}$, $f_\alpha(b) = \alpha$. Then (X_α, F_α) where $F_\alpha = \{\underline{0}, \underline{1}, f_\alpha\}$ is an fts for each $\alpha \in [0, \frac{1}{2})$. Let X be the Cartesian product of X_α for $\alpha \in [0, \frac{1}{2})$. Let F be the product fuzzy topology on X and P_α be the projection of the product X into the αth coordinate set X_α. Then $g_\alpha = P_\alpha^{-1}(f_\alpha) \in F$ for each $\alpha \in [0, \frac{1}{2})$ so that $\bigvee_\alpha g_\alpha = \frac{1}{2} \in F$. Note that here $\frac{1}{2}$ is the only inverting fuzzy subset. Hence (X, F) is type 1 invertible.

Here (X, F) is type 2 invertible even if none of the coordinate spaces are so. Conversely, the type 2 invertibility of all the coordinate spaces need not imply the type 2 invertibility of the product space as shown in the following example.

Example 6.5 Let $X_1 = \{a, b\}$ and $X_2 = \{c, d\}$. Define g_1, $g_2 \in I^{X_2}$ by $g_1(c) = \frac{2}{3}$, $g_1(d) = \frac{1}{4}$ and $g_2(c) = \frac{1}{4}$, $g_2(d) = \frac{2}{3}$. Then (X_1, F_1) and (X_2, F_2) are fuzzy topological spaces where $F_1 = \{\underline{0}, \underline{1}, \frac{2}{3}, \frac{1}{3}\}$ and $F_2 = \{\underline{0}, \underline{1}, g_1, g_2, \frac{1}{4}, \frac{2}{3}\}$. Clearly (X_1, F_1) and (X_2, F_2) are type 2 invertible. Also let (X, F) be the product fuzzy topology of (X_1, F_1) and (X_2, F_2) and let P_1 be the projection of the product X into the coordinate set X_1 and P_2 be the projection of the product X into the coordinate set X_2. Consider $f \in I^X$ defined by $f(a, c) = \frac{1}{3}$, $f(a, d) = \frac{2}{3}$, $f(b, c) =$

$\frac{1}{3}$, $f(b, d) = \frac{2}{3}$. Clearly $f = P_1^{-1}(\frac{1}{3}) \vee P_2^{-1}(g_2) \in F$. Now define $\theta : X \to X$ by $\theta(a, c) = (a, d)$, $\theta(a, d) = (a, c)$, $\theta(b, c) = (b, d)$, $\theta(b, d) = (b, c)$. Clearly θ is a homeomorphism of (X, F) and is an inverting map for f. Also identity is not an inverting map for f. Hence (X, F) is not type 2 invertible.

On the other hand, if the coordinate spaces are all invertible, then we have the following desired result:

Theorem 6.16 [2] *If the product fuzzy topology (X, F) of a family of invertible fts' (X_i, F_i), $i \in J$ is type 2 invertible, then each (X_i, F_i) is type 2 invertible.*

Proof Suppose (X, F) be type 2 invertible. If possible assume that (X_j, F_j) is not type 2 invertible for some j. Then there exists an inverting fuzzy subset g_j of (X_j, F_j) such that e is not an inverting map for g_i. Now define $\theta : X \to X$ by $\theta(x) = y$ where $y = (y_i)$ such that $y_i = x_i$, $i \neq j$, $y_j = \theta_j(x_j)$. Clearly $\theta \neq e$ is a homeomorphism of (X, F). Now consider $f = P_j^{-1}(g_j)$, clearly (f, θ) is an inverting pair of (X, F) and e is not an inverting map for f, a contradiction. Hence (X_j, F_j) is type 2 invertible. $\qquad\square$

The product of two completely invertible fuzzy topological spaces is not completely invertible as illustrated in the following example.

Example 6.6 For, let $X_1 = \{a, b, c, d\}$ and $X_2 = \{p, q, r, s\}$. Consider f_1, $g_1 \in I^{X_1}$ and f_2, $g_2 \in I^{X_2}$ defined by

$$f_1(a) = 1, \ f_1(b) = 0, \ f_1(c) = 1, \ f_1(d) = 0$$
$$g_1(a) = 0, \ g_1(b) = 1, \ g_1(c) = 0, \ g_1(d) = 1$$
$$f_2(p) = 1, \ f_2(q) = 1, \ f_2(r) = 0, \ f_2(s) = 0$$
$$g_2(p) = 0, \ g_2(q) = 0, \ g_2(r) = 1, \ g_2(s) = 1.$$

Then (X_1, F_1) and (X_2, F_2) are fuzzy topological spaces where $F_1 = \{\underline{0}, \underline{1}, f_1, g_1\}$ and $F_2 = \{\underline{0}, \underline{1}, f_2, g_2\}$. Let X be the Cartesian product of X_1 and X_2. Let F be the product fuzzy topology on X and let P_1 be the projection of the product X into the coordinate set X_1 and P_2 be the projection of the product X into the coordinate set X_2. Now consider $f \in I^X$ defined by $f = P_1^{-1}(f_1) \wedge P_2^{-1}(f_2)$. Clearly $f \in F$; $f \neq \underline{0}$ and $2|supp\ f| < |X|$. Then by Theorem 3.4, (X, F) is not invertible with respect to f so that it is not completely invertible.
But with type 2 complete invertibility, we have the following characterization.

Theorem 6.17 [2] *The product fuzzy topology (X, F) of a family of fts' (X_i, F_i), $i \in J$ is type 2 completely invertible if and only if (X_j, F_j) is type 2 completely invertible for each $j \in J$.*

Proof Let (X_i, F_i), $i \in J$ be a family of type 2 completely invertible fts. Let X be the Cartesian product $\{X_i, i \in J\}$ and let P_i be the projection of the product X into the ith coordinate set X_i. Let $\mathcal{B}$ be the defining base for the product fuzzy topology (X, F). Let $f \in \mathcal{B}$ and $f \neq \underline{0}, \underline{1}$, then $f = \bigcap_{i \in K} P_i^{-1}(g_i) : g_i \in F_i$ where K is a finite subset of J. Since $g_i \geq \frac{1}{2}$, $\forall g_i \in F_i$ and $g_i \neq \underline{0}$, $i \in J$, we have $f \geq \frac{1}{2}$ so that (f, e) is an inverting fuzzy pair of (X, F). Then by Theorem 3.9, (X, F) is type 2 completely invertible.

Conversely suppose that (X, F) is type 2 completely invertible. If possible assume that (X_j, F_j) is not type 2 completely invertible for some $j \in J$. Then by Theorem 4.10, there exists an $f_j \in F_j$ and $f_j \neq \underline{0}$ such that $f_j(x_j) < \frac{1}{2}$ for some $x_j \in X_j$. Now consider $f \in I^X$ defined by $f = P_j^{-1}(f_j)$. Clearly $f \in F$ and (f, e) is not an inverting pair of (X, F) so that (X, F) is not type 2 completely invertible, a contradiction. Hence, for each $i \in J$, (X_i, F_i) is type 2 completely invertible. □

6.6 Exercises

6.1 Prove that type 1 invertibility is an additive property.

6.2 Prove that the sum fts of a family of pairwise disjoint type 2 invertible fuzzy topological spaces is type 1 invertible.

6.3 Prove that the sum fts of a family of pairwise disjoint completely invertible fuzzy topological spaces is invertible.

6.4 Give an example of a completely invertible fuzzy topological space whose associated topological space is also completely invertible.

6.5 Prove that complete invertibility is not a hereditary property.

6.6 Let (X, F) be a type 1 completely invertible fts with $X_\lambda = \emptyset$ for $\lambda = 1$. Prove that every subspace of (X, F) is also type 1 invertible.

6.7 Let (X, F) be a type 2 completely invertible fts. Prove that any subspace (Y, F_Y) with $|F_Y| > 2$ is type 2 completely invertible.

6.8 If the associated topological space $(X, i(F))$, where $|X|$ is odd, is completely invertible then for each $g \in F$, prove that there exists a unique $\alpha \in I$ such that $g \leq \underline{\alpha}$ with $|S_\alpha(g)| \geq \frac{|X|}{2}$.

6.9 If the associated topological space $(X, i(F))$, where X is infinite, is completely invertible then for each $g \in F$, show that there exists some $\alpha \in I$ such that $S_\alpha(g)$ is infinite and $g \leq \underline{\alpha}$.

6.10 Give an example for a simple extension of a completely invertible fuzzy topological space which is completely invertible.

6.11 Give an example for a simple extension of a completely invertible fuzzy topological space which is invertible, but not completely invertible.

6.12 Show that any quotient space of a fully stratified fts is type 1 invertible.

6.13 Show that any quotient space of a topologically generated fts is type 1 invertible.

6.14 Show that the product space of a family of invertible fuzzy topological spaces is invertible.

6.15 Show that the product space of a family of type 1 invertible fuzzy topological spaces is type 1 invertible.

References

1. Mathew, S.C., Jose, A.: On the structure of invertible fuzzy topological spaces. Adv. Fuzzy Sets Syst. **40**(1), 67–79 (2010)
2. Jose, A., Mathew, S.C.: Invertibility of the related spaces of a fuzzy topological space. Thai J. Math. **13**(2), 277–289 (2015)
3. Pu, P.M., Liu, Y.M.: Fuzzy topology. 11. product and quotient spaces. J. Math. Anal. Appl. **77**, 20–37 (1980)

Chapter 7
Invertible L-Topological Spaces

This chapter extends the concept of invertibility to L-topological spaces and obtains certain properties of invertible L-topological spaces. It has been proved that stratification preserves invertibility of an L-topological space. Further, certain properties of inverting pairs are investigated. The completely invertible L-topological spaces are introduced and pinpoint some of their characteristics. Finally, we introduce two different types of invertible L-topological spaces and study their properties in relation to sums, subspaces and simple extensions. The relationship between invertibility and countability axioms in L-topological spaces is investigated. In this direction, we prove that first countable, second countable and separable properties of certain subspaces are transferable to the parent L-topological space with the help of invertibility. Further the effect of invertibility on the separation axioms in L-topological spaces is investigated and certain local to global properties of invertible L-topologies are obtained.

7.1 Invertibility in L-Topologies

This section extends the concept of invertibility to L-topological spaces and delineates its properties.

Definition 7.1 [1] An L-ts (L^X, δ) is said to be invertible with respect to a proper open L-subset v if there is an L-homeomorphism $\theta^{\rightarrow}$ of (L^X, δ) such that $\theta^{\rightarrow}(v') \leq v$. This L-homeomorphism $\theta^{\rightarrow}$ is called an inverting map for v and v is said to be an inverting L-subset of (L^X, δ).

When we say (L^X, δ) is invertible with respect to v, it is understood that $v \in \delta$ and $v \neq \underline{0}, \underline{1}$ and there exists an inverting map $\theta^{\rightarrow}$ of v. This v and $\theta^{\rightarrow}$ together are called an inverting pair of (L^X, δ). Clearly there can be different inverting pairs for an invertible L-ts.

© The Author(s), under exclusive license to Springer Nature Singapore Pte Ltd. 2022
A. Jose and S. C. Mathew, *Invertible Fuzzy Topological Spaces*,
https://doi.org/10.1007/978-981-19-3689-0_7

Example 7.1 Let $X = [-1, 0) \cup (0, 1]$ and $L = \{0, a, b, 1\}$ be the diamond-type lattice such that $a' = b$ and $b' = a$. Consider $k, l, m, n, p, q, r, s, t, u \in L^X$ defined by

$$k(x) = \begin{cases} 1; & x < 0 \\ 0; & x > 0 \end{cases} \quad l(x) = \begin{cases} 0; & x < 0 \\ 1; & x > 0 \end{cases} \quad t(x) = \begin{cases} a; & x < 0 \\ 1; & x > 0 \end{cases}$$

$$m(x) = \begin{cases} 1; & x < 0 \\ b; & x > 0 \end{cases} \quad q(x) = \begin{cases} b; & x < 0 \\ a; & x > 0 \end{cases} \quad u(x) = \begin{cases} b; & x < 0 \\ 1; & x > 0 \end{cases}$$

$$n(x) = \begin{cases} 1; & x < 0 \\ a; & x > 0 \end{cases} \quad r(x) = \begin{cases} a; & x < 0 \\ 0; & x > 0 \end{cases} \quad v(x) = \begin{cases} 0; & x < 0 \\ b; & x > 0 \end{cases}$$

$$p(x) = \begin{cases} a; & x < 0 \\ b; & x > 0 \end{cases} \quad s(x) = \begin{cases} b; & x < 0 \\ 0; & x > 0 \end{cases} \quad w(x) = \begin{cases} 0; & x < 0 \\ a; & x > 0 \end{cases}.$$

Then (L^X, δ) is an L-ts defined by $\delta = \{\underline{0}, \underline{1}, \underline{a}, \underline{b}, m, n, p, q, r, s, t, u, v, w\}$. Define $\theta : X \to X$ by $\theta(x) = -x, \forall x \in X$. Then $\theta^{\rightarrow}$ is an L-homeomorphism of (L^X, δ). Clearly $(u, \theta^{\rightarrow})$ is an inverting pair of (L^X, δ) so that (L^X, δ) is invertible.

Though the following result is evident, we record it here for the convenience of further reference.

Theorem 7.1 [1] *Let (L^X, δ) be an invertible L-ts and $v \in \delta$. Then (v, e) is an inverting pair if and only if $v' \leq v$.*

Remark 7.1 It has been proved that the stratification of any fuzzy topological space is always invertible. But, the stratification of an L-ts need not be invertible. For example, let X be the set of all natural numbers and let $L = P(Y)$ where $Y = \{a, b, c\}$. Also let $\mu = \{\underline{A} : A \in L\}$. Clearly (L^X, μ) is the stratification of the trivial L-topological space. Then for any $\underline{B} \in \mu$ we have $\underline{B'} \not\leq \underline{B}$ so that for any L-homeomorphism $\theta^{\rightarrow}$ of $(L^X, \mu), \theta^{\rightarrow}(\underline{B'}) = \underline{B'} \not\leq \underline{B}$. Hence (L^X, μ) is not invertible. However, stratification preserves the invertibility of an L-ts which is seen in the following theorem.

Theorem 7.2 [1] *The stratification of an invertible L-ts is invertible.*

Proof Let (L^X, δ) be an invertible L-ts and (L^X, μ) be the stratification of it. If $\theta^{\rightarrow}$ is a homeomorphism of (L^X, δ), then $\theta^{\rightarrow}$ is a homeomorphism of (L^X, μ). For any $u \in \mu$ is of the form either $u = v \wedge \underline{\alpha}$ or $u = w \vee \underline{\beta}$ where $v, w \in \delta$ and $\alpha, \beta \in L$. Also $\theta^{\rightarrow}(v \wedge \underline{\alpha}) = \theta^{\rightarrow}(v) \wedge \underline{\alpha}, \theta^{\leftarrow}(v \wedge \underline{\alpha}) = \theta^{\leftarrow}(v) \wedge \underline{\alpha}$ and $\theta^{\rightarrow}(w \vee \underline{\beta}) = \theta^{\rightarrow}(w) \vee \underline{\beta}$, $\theta^{\leftarrow}(w \vee \underline{\beta}) = \theta^{\leftarrow}(w) \vee \underline{\beta}$. Hence if $(u, \theta^{\rightarrow})$ is an inverting pair of (L^X, δ), it is also an inverting pair of (L^X, μ). □

Remark 7.2 Since the stratification of any fts is invertible, the converse of the above theorem is not true.

Theorem 7.3 [1] *Let (L^X, δ) be an L-ts invertible with respect to v and let $w \in \delta$ be such that $w \geq v$ and $w \neq \underline{1}$. Then (L^X, δ) is invertible with respect to w also. Moreover, there is a common inverting map for both v and w.*

Proof Let $\theta^{\rightarrow}$ be an inverting homeomorphism for v.
Then $\theta^{\rightarrow}(v') \leq v \Rightarrow v' \leq \theta^{\leftarrow}(v) \Rightarrow w' \leq \theta^{\leftarrow}(v)$
$\Rightarrow w' \leq \theta^{\leftarrow}(w) \Rightarrow \theta^{\rightarrow}(w') \leq w$. Hence (L^X, δ) is invertible with respect to w and $\theta^{\rightarrow}$ is an inverting map for w also. $\qquad\square$

Theorem 7.4 [1] *Let (L^X, δ) be an L-ts invertible with respect to v where X is finite. Then $|X| \leq 2|supp\, v|$.*

Proof (L^X, δ) is invertible with respect to v
$\Rightarrow$ there exists a homeomorphism $\theta^{\rightarrow}$ of
(L^X, δ) such that $\theta^{\rightarrow}(v') \leq v$
$\Rightarrow |supp\, \theta^{\rightarrow}(v')| \leq |supp\, v|$
$\Rightarrow |supp\, v'| \leq |supp\, v|$
$\Rightarrow |X| - |supp\, v| \leq |supp\, v|$
$\Rightarrow |X| \leq 2|supp\, v|$. $\qquad\square$

Theorem 7.5 [1] *Let $\theta^{\rightarrow}$ be a homeomorphism of an L-ts (L^X, δ) and $v \in \delta$; $v \neq \underline{0}, \underline{1}$. Then the invertibility of (L^X, δ) with respect to the following pairs is equivalent.*

(i) $(v, \theta^{\rightarrow})$;
(ii) $(v, \theta^{\leftarrow})$;
(iii) $(\theta^{\leftarrow}(v), \theta^{\leftarrow})$;
(iv) $(\theta^{\rightarrow}(v), \theta^{\rightarrow})$;
(v) $(\theta^{\rightarrow}(v), \theta^{\leftarrow})$;
(vi) $(\theta^{\leftarrow}(v), \theta^{\rightarrow})$.

Proof The assertions of the theorem follow from the following set of inequalities:
$\theta^{\rightarrow}(v') \leq v \Rightarrow v' \leq \theta^{\rightarrow}(v) \Rightarrow \theta^{\leftarrow}(v') \leq v$
$\Rightarrow (\theta^{\leftarrow}(v))' \leq v \Rightarrow \theta^{\leftarrow}(\theta^{\leftarrow}(v))' \leq \theta^{\leftarrow}(v)$
$\Rightarrow (\theta^{\leftarrow}(v))' \leq v \Rightarrow v' \leq \theta^{\rightarrow}(v) \Rightarrow \theta^{\rightarrow}(\theta^{\rightarrow}(v))' \leq \theta^{\rightarrow}(v)$
$\Rightarrow (\theta^{\rightarrow}(v))' \leq \theta^{\rightarrow}(\theta^{\rightarrow}(v)) \Rightarrow \theta^{\leftarrow}(\theta^{\rightarrow}(v))' \leq \theta^{\rightarrow}(v)$
$\Rightarrow (\theta^{\rightarrow}(v))' \leq v \Rightarrow \theta^{\rightarrow}(\theta^{\leftarrow}(v))' \leq \theta^{\leftarrow}(v)$
$\Rightarrow (\theta^{\rightarrow}(v))' \leq v \Rightarrow \theta^{\rightarrow}(v') \leq v$. $\qquad\square$

Corollary 7.1 [1] *Let (L^X, δ) be an invertible L-ts and (v, θ) be an inverting pair of (L^X, δ). Then $v' \leq \theta^{\rightarrow}(v) \wedge \theta^{\leftarrow}(v)$.*

Proof Since $(v, \theta^{\rightarrow})$ is an inverting pair of (L^X, δ), $v' \leq \theta^{\leftarrow}(v)$. By above theorem $(v, \theta^{\leftarrow})$ is also an inverting pair of (L^X, δ) so that $v' \leq \theta^{\rightarrow}(v)$. $\qquad\square$

Theorem 7.6 [1] *Let (L^X, δ) be an invertible L-ts and $(v, \theta^{\rightarrow})$ be an inverting pair of (L^X, δ). Then the following statements are equivalent:*
(i) v and $\theta^{\rightarrow}(v)$ are not quasi-coincident.
(ii) $\theta^{\rightarrow}(v) = v'$.
(iii) $\theta^{\leftarrow}(v) = v'$.

Proof (i) $\iff$ (ii) By Theorem 7.5, $(v, \theta^{\leftarrow})$ is an inverting pair of (L^X, δ) so that $\theta^{\leftarrow}(v') \leq v \Rightarrow v' \leq \theta^{\rightarrow}(v)$. Suppose that v and $\theta^{\rightarrow}(v)$ are not quasi-coincident. Then $\theta^{\rightarrow}(v) \leq v'$ so that $\theta^{\rightarrow}(v) = v'$. Conversely suppose that $\theta^{\rightarrow}(v) = v' \Rightarrow \theta^{\rightarrow}(v) \leq v'$. Hence, v and $\theta^{\rightarrow}(v)$ are not quasi-coincident. Consequently, $\theta^{\rightarrow}(v) = v'$ if and only if v and $\theta^{\rightarrow}(v)$ are not quasi-coincident.
(ii) $\iff$ (iii) $\theta^{\rightarrow}(v) = v'$ if and only if $v = \theta^{\leftarrow}(v')$ if and only if $v' = \theta^{\leftarrow}(v)$. □

Remark 7.3 From the above theorem it follows that if v and $\theta^{\rightarrow}(v)$ are not quasi-coincident, then $\theta^{\rightarrow}(v) = \theta^{\leftarrow}(v)$ and v is closed in (L^X, δ).

Theorem 7.7 [1] *Let $(v, \theta^{\rightarrow})$ be an inverting pair of (L^X, δ) such that v and $\theta^{\rightarrow}(v)$ are not quasi-coincident. If $|S_\alpha(v)| \leq 1$, $\forall \alpha \in L$, then $\theta^2 = e$ and $\theta^{\rightarrow}$ is the only inverting map.*

Proof By Theorem 7.6, v and $\theta^{\rightarrow}(v)$ are not quasi-coincident if and only if $\theta^{\rightarrow}(v) = v' = \theta^{\leftarrow}(v)$. But we claim that $\theta = \theta^{-1}$. If $\theta \neq \theta^{-1}$, then there exists at least one $x \in X$ such that $\theta(x) \neq \theta^{-1}(x)$. Let $\theta(x) = y$ and $\theta^{-1}(x) = z$ where $y \neq z$. Then $v(z) = v(\theta^{-1}(x)) = \theta^{\rightarrow}(v)(x) = \theta^{\leftarrow}(v)(x) = v(\theta(x)) = v(y)$, a contradiction. Hence the claim so that $\theta^2 = e$.
Now if possible let $\eta^{\rightarrow}$ be an inverting map of v other than $\theta^{\rightarrow}$. Then by Corollary 7.1, $v' \leq \eta^{\rightarrow}(v) \Rightarrow \theta^{\rightarrow}(v) \leq \eta^{\rightarrow}(v)$. We claim that $\theta^{\rightarrow}(v) = \eta^{\rightarrow}(v)$. If $\theta^{\rightarrow}(v)(x) < \eta^{\rightarrow}(v)(x)$ for some $x \in X$, then $\eta^{\rightarrow}(v)'(x) < \theta^{\rightarrow}(v)'(x)$. But $v'(x) = \theta^{\rightarrow}(v)(x) = v(y)$ for some $y \in X$. Then $\eta^{\rightarrow}(v)(y) = \eta^{\rightarrow}(v')(x) = \eta^{\rightarrow}(v)'(x) < \theta^{\rightarrow}(v)'(x) = \theta(v')(x) = \theta(v)(y)$, a contradiction. Hence the claim. Since $|S_\alpha(v)| \leq 1$ for all $\alpha \in L$, we have $v(a) \neq v(b)$ for every distinct $a, b \in X$. Now we have to prove that $\theta^{\rightarrow} = \eta^{\rightarrow}$. If possible assume that $\theta^{\rightarrow} \neq \eta^{\rightarrow}$. Then there exists an $x \in X$ such that $\theta^{-1}(x) \neq \eta^{-1}(x)$. Let $y = \theta^{-1}(x)$ and $z = \eta^{-1}(x)$. Now $v(y) = \theta^{\rightarrow}(v)(x) = \eta^{\rightarrow}(v)(x) = v(z)$, a contradiction. Thus $\theta^{\rightarrow}$ is the only inverting map for v. □

Remark 7.4 Converse of the above theorem is not true. For example, let $X = \{2, 3, 4, 5\}$ and L be the following lattice.

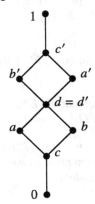

Consider $p, q, m, n, u, v, w \in L^X$ defined by

$$u(2) = a, \ u(3) = a', \ u(4) = a', \ u(5) = a$$
$$v(2) = a', \ v(3) = a, \ \ v(4) = a, \ v(5) = a'$$

$$w(2) = d, \ w(3) = c, \ \ w(4) = d, \ w(5) = c$$
$$m(2) = a, \ m(3) = c, \ \ m(4) = d, \ m(5) = c$$
$$n(2) = d, \ n(3) = c, \ \ n(4) = a, \ n(5) = c$$
$$p(2) = d, \ p(3) = a', \ p(4) = a', \ p(5) = a$$
$$q(2) = a', \ q(3) = a, \ \ q(4) = d, \ q(5) = a'.$$

Then (L^X, δ) is an L-ts where $\delta = \{\underline{0}, \underline{1}, p, q, m, n, u, v, w, \underline{a}, \underline{a'}\}$. Define $\theta : X \to X$ by $\theta(2) = 4$, $\theta(3) = 5$, $\theta(4) = 2$, $\theta(5) = 3$. Clearly $\theta^\to$ is an inverting map for v in (L^X, δ) and v and $\theta^\to(v)$ are not quasi-coincident. Here $\theta^2 = e$ and $\theta^\to$ is the only inverting map for v, but $v(5) = v(2)$.

Remark 7.5 In Chap. 3, it is shown that, if an fts is invertible with respect to an open fuzzy subset not containing any crisp singleton, then identity is an inverting map of the fts. But for an L-ts this need not be true. For example, let $X = \{x \in \mathbb{N} : 1 \le x \le 100\}$ and $L = P(Y)$ where $Y = \{1, 2, 3, 4\}$. Let $A = \{1, 2\}$ and $B = \{3, 4\}$. Now consider $u, v \in L^X$ defined by

$$u(x) = \begin{cases} A; \ 1 \le x \le 50 \\ B; \ 51 \le x \le 100 \end{cases}$$

$$v(x) = \begin{cases} B; \ 1 \le x \le 50 \\ A; \ 51 \le x \le 100 \end{cases}.$$

Then (L^X, δ) is an L-ts where $\delta = \{\underline{0}, \underline{1}, u, v\}$. Clearly (L^X, δ) is invertible with respect to u, but u does not contain any crisp singleton. Note that identity is not an inverting map of (L^X, δ).

7.2 Completely Invertible L-Topological Spaces

This section introduces completely invertible L-topological spaces and derives certain characterizations of them. Also the effect of complete invertibility on the connectedness of an L-topological space is investigated.

Definition 7.2 [1] A non-trivial L-ts (L^X, δ) is said to be completely invertible if for any $v \in \delta$ and $v \neq \underline{0}$, there is a homeomorphism $\theta^\to$ of (L^X, δ) such that $\theta^\to(v') \le v$.

It should be noted that for a completely invertible L-ts every non-empty open L-subset is an inverting L-subset.

We continue with the following easy consequence of the aforesaid definition.

Theorem 7.8 [1] *Any L-ts (L^X, δ) containing an open L-point with $|X| > 2$ is not completely invertible.*

Theorem 7.9 [1] *Let (L^X, δ) be an L-ts with δ_0 as a base. Then (L^X, δ) is completely invertible if and only if (L^X, δ) is invertible with respect to all members of δ_0.*

Proof The necessary part is obvious. Conversely suppose (L^X, δ) be invertible with respect to v for all $v \in \delta_0$. Let u be any proper open L-subset of X. Then u is of the form $u = \vee_{v \in \delta_u} v$ where $\delta_u \subset \delta_0$. Now by Theorem 7.3, (L^X, δ) is invertible with respect to u also. $\square$

Remark 7.6 We know that an fts (X, F) is completely invertible if and only if for each proper closed fuzzy subset h and each $g \in F$ and $g \neq \underline{0}$, there is a homeomorphism θ of (X, F) such that $\theta(h) \leq g$. But in the case of an L-ts, this doesn't hold. For example, consider the lattice given below.

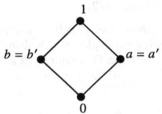

Let X be the set of all real numbers. Then the L-ts (L^X, δ) where $\delta = \{\underline{0}, \underline{1}, \underline{a}, \underline{b}\}$ is completely invertible. Here $\underline{a}$ is a closed L-subset of (L^X, δ), but there exists no homeomorphism $\theta^\rightarrow$ of (L^X, δ) such that $\theta^\rightarrow(\underline{a}) \leq \underline{b}$.

Theorem 7.10 [1] *Let (L^X, δ) be an L-ts invertible with respect to $v = \bigwedge_{u \in \delta - \{\underline{0}\}} u$.*

Then (L^X, δ) is completely invertible.

Proof Let (L^X, δ) be an L-ts invertible with respect to $v = \bigwedge_{u \in \delta - \{\underline{0}\}} u$. Clearly $v \leq u$ for all $u \in \delta$; $u \neq \underline{0}$ and then by Theorem 7.3, (L^X, δ) is invertible with respect to every $u \in \delta$ and $u \neq \underline{0}, \underline{1}$. $\square$

Definition 7.3 [1] Let (L^X, δ) be an L-ts. Then it is called a c-L-ts if the members of δ are characteristic functions.

The following theorem gives a characterization for completely invertible finite c-L-ts.

Theorem 7.11 [1] *A finite c-L-ts (L^X, δ) is completely invertible if and only if $|X|$ is even and $\delta = \{\underline{0}, \underline{1}, u, u'\}$.*

Proof Let $(u, \theta^\rightarrow)$ be an inverting pair of (L^X, δ). Then by Theorem 7.4, $|supp\ u| \geq \frac{|X|}{2}$. Let $v_0 = u$ and $\theta_0^\rightarrow = \theta^\rightarrow$. Define $v_1 = v_0 \wedge \theta_0^\rightarrow(v_0)$. Clearly $v_1 \in \delta$ and $|supp\ v_1| < |supp\ v_0|$. Let $\theta_1^\rightarrow$ be an inverting map of v_1. Let $v_2 = v_1 \wedge \theta_1^\rightarrow(v_1)$.

Clearly $v_2 \in \delta$ and $|supp\ v_2| < |supp\ v_1|$. Continuing like this, after a certain stage there exists a $v_n \in \delta$ and $v_n \neq \underline{0}$, where $n \in \mathbb{N}$ and $v_n = v_{n-1} \wedge \theta_{n-1}^{\rightarrow}(v_{n-1})$ such that $|supp\ v_n| \leq \frac{|X|}{2}$. Since (L^X, δ) is completely invertible, $|supp\ v_n| = \frac{|X|}{2}$. Now consider $k = \theta_n^{\rightarrow}(v_n) \wedge u \in \delta$, then $|supp\ k| < \frac{|X|}{2}$ so that by Theorem 7.4, $k = \underline{0}$. Since $\theta_n^{\rightarrow}(v_n) = v_n'$, $k = \underline{0} \Rightarrow v_n = u$. Hence $|supp\ u| = \frac{|X|}{2}$. Thus $u \in \delta$ and $u \neq \underline{0} \Rightarrow |supp\ u| = \frac{|X|}{2}$.

Now if possible let $v \in \delta$ and $v \neq u$. Then from above we have $|supp\ v| = \frac{|X|}{2}$. Let $w = v \wedge u$, then $w \in \delta$ and $|supp\ w| < \frac{|X|}{2}$. Since (L^X, δ) is completely invertible, $w = \underline{0}$ which implies that $v = u'$. Hence $\delta = \{\underline{0}, \underline{1}, u, u'\}$.

Conversely assume that $|X|$ is even and $\delta = \{\underline{0}, \underline{1}, u, u'\}$. Let $supp\ u = \{x_1, x_2, ..., x_n\}$ and $supp\ u' = \{y_1, y_2, ..., y_n\}$. Define $\theta : X \to X$ by $\theta(x_i) = y_i$ and $\theta(y_i) = x_i$ where $i = 1, 2, ..., n$. Clearly $(u, \theta^{\rightarrow})$ and $(u', \theta^{\rightarrow})$ are inverting pairs of (L^X, δ). Consequently (L^X, δ) is completely invertible. $\square$

Definition 7.4 [2] Let (L^X, δ) be an L-ts and $u, v \in L^X$, u and v are called separated, if $u^- \wedge v = u \wedge v^- = \underline{0}$. An L-subset w is called connected, if there does not exist separated $u, v \in L^X - \{\underline{0}\}$ such that $w = v \vee u$. (L^X, δ) is connected if $\underline{1}$ is connected.

Theorem 7.12 [2] *Let (L^X, δ) be an L-ts, $A \in L^X$. Then the following conditions are equivalent:*

(i) A is connected.
(ii) $u, v \in L^X$ are separated and $A \leq (u \vee v) \Rightarrow A \leq u$ or $A \leq v$.

Theorem 7.13 [2] *An L-ts (L^X, δ) is connected if and only if $u, v \in \delta$, $u \vee v = \underline{1}$, $u \wedge v = \underline{0} \Rightarrow u = \underline{0}$ or $v = \underline{0}$.*

Definition 7.5 [2] Let (L^X, δ) be an L-ts and $u \in L^X$, u is called a connected component of (L^X, δ), if u is a maximally connected L-subset of (L^X, δ), i.e. $v \in L^X$, v is connected and $v \geq u \Rightarrow v = u$.

Theorem 7.14 [2] *Let (L^X, δ), (L^Y, μ) be L-fts' and $\theta^{\rightarrow} : (L^X, \delta) \to (L^Y, \mu)$ be a continuous mapping. If $u \in L^X$ is connected. Then $\theta^{\rightarrow}(u)$ is connected.*

Proof Let (L^X, δ) be a L-connected L-ts and $\theta^{\rightarrow} : L^X \to L^Y$ be a continuous onto function from (L^X, δ) to (L^Y, μ). If possible assume that (L^Y, μ) is not L-connected. Then there exists two non-empty open L-subsets u and v such that $u \vee v = \underline{1}$. Hence $\theta^{\leftarrow}(u)$ and $\theta^{\leftarrow}(v)$ are two non-empty open L-subsets of X such that $\theta^{\leftarrow}(u) \vee \theta^{\leftarrow}(v) = \underline{1}$ so that (L^X, δ) is not L-connected, a contradiction. $\square$

Theorem 7.15 [1] *Let (L^X, δ) be a completely invertible L-ts containing a connected open crisp subset A. If (L^X, δ) is not connected, then A and A' are the connected components of (L^X, δ) and they are homeomorphic.*

Proof Let $\theta^{\rightarrow}$ be an inverting homeomorphism for A. Then by Theorem 7.5, $A' \leq \theta^{\rightarrow}(A)$. If (L^X, δ) is not connected, there exists two separated L-subsets $u, v \in \delta$ and $u, v \neq \underline{0}$ such that $u \vee v = \underline{1}$. Since A is connected, by Theorem 7.12, $A \leq u$ or $A \leq v$. Since $\theta^{\rightarrow}(A)$ is also connected, again by Theorem 7.12, we get $\theta^{\rightarrow}(A) \leq u$ or $\theta^{\rightarrow}(A) \leq v$. Without loss of generality, assume that $A \leq u$. If $\theta^{\rightarrow}(A) \leq u$, then $A \vee \theta^{\rightarrow}(A) \leq u$. But since $A' \leq \theta^{\rightarrow}(A)$, $A \vee \theta^{\rightarrow}(A) = \underline{1}$ so that $u = \underline{1}$. Then since u and v are separated, $u^- \wedge v = \underline{0} \Rightarrow v = \underline{0}$, a contradiction. Hence $\theta^{\rightarrow}(A) \leq v$ so that $A \wedge \theta^{\rightarrow}(A) = \underline{0}$ and $A' = \theta^{\rightarrow}(A)$. Consequently A and A' are the connected components of (L^X, δ) and they are homeomorphic.   □

Theorem 7.16 [1] *If a typically T_1 L-ts (L^X, δ) with a non-empty, open connected crisp subset A is completely invertible, then (L^X, δ) is connected.*

Proof: If (L^X, δ) is not connected, then by Theorem 7.15, A and A' are the connected components of (L^X, δ) and they are homeomorphic. Let $x \in A$, then $B = A \wedge x'$ is open. Let $\theta^{\rightarrow}$ be an inverting homeomorphism for B. Then $\theta^{\rightarrow}(B') \leq B$ so that $\theta^{\rightarrow}(A') \leq B$. But then $\theta^{\rightarrow}(A')$ is a component of (L^X, δ) properly contained in the connected component A, a contradiction.   □

7.3 Types of Invertible L-Topologies

Based on the inverting maps two types of invertible L-topological spaces are introduced. The sums, subspaces and simple extensions of these spaces are also studied.

Definition 7.6 [1] An invertible L-ts (L^X, δ) is said to be type 1 if identity is an inverting map.

Definition 7.7 [1] An invertible L-ts (L^X, δ) is said to be type 2 if identity is an inverting map for all the inverting L-subsets.

Remark 7.7 Clearly every type 2 invertible L-ts is type 1 invertible, but not vice versa. For a given non-empty set X, there always exists type 2 invertible fuzzy topologies on X. But this is no longer true in the case of an L-topology. For example, if $L = P(Y)$ for some non-empty set Y, then there does not exist an inverting L-subset of X with identity as an inverting map.

As a consequence of Theorem 7.1, we get the following two theorems:

Theorem 7.17 [1] *An L-ts (L^X, δ) is type 1 invertible if and only if there exists a $u \in \delta$ and $u \neq \underline{1}$ such that $u' \leq u$.*

Theorem 7.18 [1] *An L-ts (L^X, δ) is type 2 completely invertible if and only if $v' \leq v$ for every $v \in \delta$ and $v \neq \underline{0}$.*

Theorem 7.19 [1] *Let (L^X, δ) be an invertible L-ts which is not type 1. Then (L^X, δ) has at least two inverting L-subsets.*

Proof Let $(v, \theta^{\rightarrow})$ be an inverting pair of (L^X, δ). Then by Theorem 7.5, $\theta^{\rightarrow}(v)$ is an inverting L-subset of (L^X, δ). Also by Theorem 7.1, there exists an $x \in X$ such that $v'(x) \nleq v(x)$. But by Corollary 7.1, $v'(x) \leq \theta^{\rightarrow}(v)(x)$ so that $v \neq \theta^{\rightarrow}(v)$. $\qquad\square$

Remark 7.8 In Chap. 4, it has been proved that every completely invertible fts (X, F) where X is finite is either type 2 or a c-fts. But this is not true for L-topological spaces. For instance, the L-ts given in Example 7.5 is completely invertible, but is neither a c-L-ts nor a type 2 invertible L-ts.

Definition 7.8 [2] Let $\{(L^{X_t}, \delta_t) : t \in T\}$ be a family of L-ts', different X_t's be disjoint and $X = \cup_{t \in T} X_t$. Define the sum topology of $\{\delta_t : t \in T\}$ on L^X, denoted by $\oplus_{t \in T} \delta_t$, as follows: $\forall u \in L^X$, $u \in \oplus_{t \in T} \delta_t \Longleftrightarrow \forall t \in T, u|_{X_t} \in \delta_t$. The L-ts $(L^X, \oplus_{t \in T} \delta_t)$ is called the sum space of $\{(L^{X_t}, \delta_t) : t \in T\}$.

Theorem 7.20 [1] *The sum space of a family of invertible L-topological spaces is invertible.*

Proof Let (L^{X_t}, δ_t), $t \in T$ be a family of pairwise disjoint invertible L-topological spaces. For each $t \in T$, let v_t be an inverting subset for (L^{X_t}, δ_t). Let (L^X, δ) be the sum L-ts. Define $v \in L^X$ by $v(x) = v_t(x); \; x \in X_t$. Clearly $v \in \delta$. Since (L^{X_t}, δ_t) is invertible with respect to v_t, there exists a homeomorphism $\theta_t^{\rightarrow}$ of (L^{X_t}, δ_t) such that $\theta_t^{\rightarrow}(v_t)' \leq v_t$. Now define $\theta : X \to X$ by $\theta(x) = \theta_t(x)$, $x \in X_t$. Then clearly $\theta^{\rightarrow}$ is a homeomorphism of (L^X, δ). Let x be any element in X. Then $x \in X_t$ for some $t \in T$ so that $\theta^{\rightarrow}(v')(x) = \theta_t^{\rightarrow}(v_t)'(x) \leq v_t(x) = v(x)$. Since x is arbitrary, we get $\theta^{\rightarrow}(v') \leq v$. Consequently (L^X, δ) is invertible with respect to v. $\qquad\square$

Remark 7.9 The sum space of a family of pairwise disjoint type 2 invertible L-topological spaces is invertible but need not be type 2. Also the sum space of a family of completely invertible L-ts need not be completely invertible.

Definition 7.9 [2] Let (L^X, δ) be an L-ts, $Y \subset X, Y \neq \emptyset$. Then the relative topology of δ on Y is given by $\delta|_Y = \{u|_Y : u \in \delta\}$ and the pair $(L^Y, \delta|_Y)$ is called an L-subspace of (L^X, δ), or a subspace for short.

Definition 7.10 [2] An L-topological property $\mathcal{P}$ is called hereditary, if for each subspace of an L-ts with property $\mathcal{P}$ also has property $\mathcal{P}$.

Remark 7.10 Type 2 invertibility, type 1 invertibility and invertibility are not hereditary properties. The following theorem gives a subclass of invertible L-topological spaces for which type 1 invertibility is a hereditary property.

Theorem 7.21 [1] *Let (L^X, δ) be a type 1 invertible L-ts with $X_{\lambda} = \emptyset$ for $\lambda = 1$. Then every L-subspace of (L^X, δ) is also type 1 invertible.*

Proof Let (L^X, δ) be a type 1 invertible L-ts and Y be a non-empty subset of X. By Theorem 7.17, there exists a $u \in \delta$; $u \neq \underline{1}$ such that $u' \leq u$. Clearly $v = u \wedge Y$ is a proper open L-subset of the L-subspace $(L^Y, \delta|_Y)$ and $v' \leq v$. Then identity is an inverting map for v so that $(L^Y, \delta|_Y)$ is type 1 invertible. $\qquad\square$

Theorem 7.22 [1] *Let* (L^X, δ) *be a type 2 completely invertible L-ts. Then any subspace* (Y, F_Y) *with* $|F_Y| > 2$ *is type 2 completely invertible.*

Proof Follows from Theorem 7.18.   □

Definition 7.11 [3] Let (L^X, δ) be an *L*-ts and suppose that $v \in L^X$ and $v \notin \delta$. Then the collection $\delta(v) = \{v_1 \vee (v_2 \wedge v) : v_1, v_2 \in \delta\}$ is called the simple extension of δ determined by v.

Theorem 7.23 [1] *The simple extension of a type 1 invertible L-ts is type 1 invertible.*

Proof Let (L^X, δ) be a type 1 invertible *L*-ts and $(L^X, \delta(v))$ be the simple extension of δ determined by v where $v \in L^X$ and $v \notin \delta$. Let (u, e) be an inverting pair of (L^X, δ). Since $\delta \subset \delta(v)$, (u, e) is also an inverting pair of $(L^X, \delta(v))$ so that $(L^X, \delta(v))$ is type 1 invertible.   □

Corollary 7.2 [1] *The simple extension of a type 2 invertible L-ts is type 1 invertible.*

Proof Since every type 2 invertible *L*-ts is type 1 invertible, the result follows from Theorem 6.3.   □

Remark 7.11 The simple extension of a type 2 invertible *L*-ts need not be type 2 invertible and the simple extension of a type 2 completely invertible *L*-ts need not be completely invertible. Above all, the simple extension of a completely invertible *L*-ts need not even be invertible. These are shown in Chap. 6. Further in Chap. 6, it has been proved that if (X, F) is a type 2 completely invertible fts and $g \in I^X$ such that $g \notin F$, then $(X, F(g))$ is type 2 completely invertible if and only if $g \geq g'$. But this is no longer true in the case of *L*-topological spaces. For example, let X be the set of all non-zero integers and L be the lattice given in remark 7.6. Now consider $u, v \in L^X$ defined by

$$u(x) = \begin{cases} a; & x < 0 \\ b; & x > 0 \end{cases}$$

$$v(x) = \begin{cases} b; & x < 0 \\ a; & x > 0 \end{cases}$$

Then (L^X, δ) where $\delta = \{\underline{0}, \underline{1}, u, v\}$ is type 2 completely invertible. Now consider the simple extension $\delta(\underline{a})$ of δ. Clearly the *L*-subset w of X defined by

$$w(x) = \begin{cases} a; & x < 0 \\ 0; & x > 0 \end{cases}$$

is open in $(L^X, \delta(\underline{a}))$, but is not an inverting *L*-subset of $(L^X, \delta(\underline{a}))$.
However, the following theorem gives a sufficient condition for the desired result.

Theorem 7.24 [1] *Let* (L^X, δ) *be a type 2 completely invertible L-ts and* $v \in L^X$ *be such that* $v \notin \delta$ *and* $v \geq u$, $\forall u \in \delta$ *and* $u \neq \underline{1}$. *Then* $(L^X, \delta(v))$ *is type 2 completely invertible.*

Proof Since (L^X, δ) is type 2 completely invertible, by Theorem 7.18, we have $u' \leq u, \forall u \neq \underline{0} \in \delta$. Now consider $k \in \delta(v); k \neq \underline{0}$. Then $k = p \vee (q \wedge v)$ for some $p, q \in \delta$. If $p \neq \underline{0}$, then $k' \leq p' \leq p \leq k$. If $p = \underline{0}$, then $k = q \wedge v$. Assume that $q \neq \underline{1}$, then since $v \geq q$, $k = q$ so that $k' = q' \leq q = k$. If $q = \underline{1}$, then $k = v$. Let u be any proper open L-subset in (L^X, δ), then $u \leq v \Rightarrow v' \leq u'$. Since $u' \leq u$ this implies that $v' \leq v$ so that $k' \leq k$. Thus, in all the possible cases, we have $k' \leq k$ for every $k \in \delta(v)$ and $k \neq \underline{0}$. Consequently (L^X, δ) is type 2 completely invertible, as required. $\qquad\square$

Theorem 7.25 [1] *Let (L^X, δ) be a type 2 completely invertible L-ts and $v \in L^X$ such that $v \notin \delta$. Then v is an inverting L-subset of $(L^X, \delta(v))$ if and only if $v \geq v'$.*

Proof If $v \geq v'$, then by Theorem 7.1, v is an inverting L-subset of $(L^X, \delta(v))$. Conversely suppose that v is an inverting L-subset of (L^X, δ). If possible assume that there exists an $x \in X$ such that $v'(x) \not\leq v(x)$. Let $(v, \theta^{\rightarrow})$ be an inverting pair of $(L^X, \delta(v))$, then by Corollary 7.1, $v'(x) \leq \theta^{\rightarrow}(v)(x)$. But $\theta^{\rightarrow}(v) = p \vee (q \wedge v)$ for some $p, q \in \delta$. If $p \neq \underline{0}$, then $\theta^{\rightarrow}(v)' \leq p' \leq p \leq \theta^{\rightarrow}(v) \Rightarrow v' \leq v$, a contradiction. Hence $p = \underline{0}$, so that $\theta(v) = q \wedge v$. Now $v'(x \leq \theta^{\rightarrow}(v)(x) \Rightarrow v'(x) \leq (q \wedge v)(x) \Rightarrow v'(x) \leq v(x)$, a contradiction. Consequently $v' \leq v$. $\qquad\square$

Remark 7.12 It has been shown that the simple extension of a type 2 completely invertible fts is type 2 invertible. But this is not true in the case of L-topological spaces. For example, let X be the set of all non-zero real numbers and L be the lattice given below.

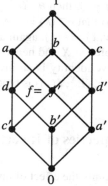

Now consider $p, q \in L^X$, defined by

$$p(x) = \begin{cases} a; \ x < 0 \\ c; \ x > 0 \end{cases}$$

$$q(x) = \begin{cases} c; \ x < 0 \\ a; \ x > 0 \end{cases}.$$

Then (L^X, δ) where $\delta = \{\underline{0}, \underline{1}, p, q, \underline{f}\}$ is a type 2 completely invertible L-ts. Now consider the simple extension $\delta(\underline{b})$ of δ. Clearly the L-subset k of X defined by

$$k(x) = \begin{cases} d; & x < 0 \\ d'; & x > 0 \end{cases}$$

is open in $(L^X, \delta(\underline{b}))$ and identity is not an inverting map for k. Define $\theta : X \to X$ by $\theta(x) = -x$, $\forall x \in X$. Then $\theta^{\to}$ is a homeomorphism of $(L^X, \delta(\underline{b}))$ and $(k, \theta^{\to})$ is an inverting pair of $(L^X, \delta(\underline{b}))$. Consequently $(L^X, \delta(\underline{b}))$ is invertible, but not type 2.

However, if L is a chain, the type 2 nature will be retained with simple extensions in the case of a type 2 completely invertible L-ts as seen in the following theorem.

Theorem 7.26 [1] *Let L be a chain and (L^X, δ) be a type 2 completely invertible L-ts. Then every simple extension of (L^X, δ) is type 2 invertible.*

Proof Let $(L^X, \delta(v))$ be the simple extension of (L^X, δ) determined by $v \in L^X$ and $v \notin \delta$. Since (L^X, δ) is type 2 completely invertible, by Theorem 7.18 we have $u' \le u$, $\forall u \ne \underline{0} \in \delta$. Now by Theorem 6.3, $(L^X, \delta(v))$ is type 1 invertible. Let k be an inverting L-subset of $(L^X, \delta(v))$ with inverting map $\theta^{\to}$. Then we have $k = p \vee (q \wedge v)$, for some $p, q \in \delta$.

Further we claim that $k' \le k$. If $p \ne \underline{0}$, then $k' \le p' \le p \le k$, as claimed. If $p = \underline{0}$, then $k = q \wedge v$ and $q \ne \underline{0}$. If possible assume that there exists an $x \in X$ such that $k'(x) \not\le k(x)$. Since L is a chain, there arise two cases:

Case 1: $q(x) \le v(x)$. Then $k'(x) = q'(x) \le q(x) = k(x)$, a contradiction.

Case 2: $v(x) < q(x)$. Then $k(x) = v(x)$. Since $(k, \theta^{\to})$ is an inverting pair of $(L^X, \delta(v))$, then by Corollary 7.1, we have $k'(x) \le \theta^{\to}(k)(x)$. But $\theta^{\to}(k) = m \vee (n \wedge v)$ for some $m, n \in \delta$. If $m \ne \underline{0}$, then $\theta^{\to}(k)' \le m' \le m \le \theta^{\to}(k) \Rightarrow k' \le k$, a contradiction. Hence $m = \underline{0}$ so that $\theta^{\to}(k) = n \wedge v$. Now $k'(x) \le \theta^{\to}(k)(x) \Rightarrow k'(x) \le (n \wedge v)(x) \Rightarrow k'(x) \le v(x) = k(x)$, again a contradiction.

Consequently $k'(x) \le k(x)$ for all $x \in X$ and hence the claim.

Since k is arbitrary, now we have identity is an inverting map for every inverting L-subset of $(L^X, \delta(v))$. Therefore, $(L^X, \delta(v))$ is type 2 invertible. $\square$

7.4 Local to Global Properties of Invertible L-Topologies

In this section, we investigate mainly the effect of invertibility on separation axioms in L-topological spaces and obtain certain sufficient conditions for an invertible L-topological space to satisfy them. The relationship between invertibility and countability axioms in L-topological spaces is also examined.

Definition 7.12 [2] Let (L^X, δ) be an L-ts, $\delta_0 \subset \delta$. δ_0 is called a base of δ, if $\delta = \{\vee \mathcal{A} : \mathcal{A} \subset \delta_0\}$.

Definition 7.13 [2] Let (L^X, δ) be an L-ts and $A \in L^X$. Then the weight $w(\delta)$ of (L^X, δ) is defined by $w(\delta) = \min\{|\mathcal{B}| : \mathcal{B}$ is a base of $\delta\}$.

Definition 7.14 [2] Let (L^X, δ) be an L-ts, (L^X, δ) is called second countable or C_{II}, if $w(\delta) \leq \omega$.

Theorem 7.27 [4] *Let (L^X, δ) be an L-ts invertible with respect to an open crisp subset Y. If $(L^Y, \delta|_Y)$ is second countable, then (L^X, δ) is also second countable.*

Proof Let $\theta^\rightarrow$ be an inverting homeomorphism for Y. Then by Theorem 7.1, $Y' \leq \theta^\rightarrow(Y)$. Consequently any $u \in \delta$ is of the form $u = u \wedge (Y \vee \theta^\rightarrow(Y)) = (u \wedge Y) \vee (u \wedge \theta^\rightarrow(Y)) = v_1 \vee v_2$ where $v_1 = u \wedge Y \in \delta|_Y$ and $v_2 = u \wedge \theta^\rightarrow(Y) \in \delta|_{\theta^\rightarrow(Y)}$. Let $\mathcal{B}$ be a base for $(L^Y, \delta|_Y)$ with $|\mathcal{B}| = w(\delta|_Y)$. Since $(L^Y, \delta|_Y)$ is second countable, $|\mathcal{B}| \leq \omega$. Now $v_1 = \vee\mathcal{A}$ where $\mathcal{A} \subset \mathcal{B}$ and $v_2 = \vee\theta^\rightarrow(\mathcal{D})$ where $\mathcal{D} \subset \mathcal{B}$. Hence $\delta_0 = \mathcal{B} \cup \theta^\rightarrow(\mathcal{B})$ is a base for (L^X, δ) and $|\delta_0| \leq \omega$. $\qquad \square$

Definition 7.15 [2] Let (L^X, δ) be an L-ts and $x_a \in Pt(L^X)$. Then $u \in \delta$ is called a quasi-coincident neighbourhood (Q-neighbourhood) of x_a in (L^X, δ), if x_a quasi-coincides with u. The family of all the Q-neighbourhoods of x_a in (L^X, δ) is called the Q-neighbourhood system of x_a, denoted by $\mathcal{Q}(x_a)$.

Definition 7.16 [2] Let (L^X, δ) be an L-ts and $x_\lambda \in Pt(L^X)$. A subfamily $\mathcal{A} \subset \mathcal{Q}(x_\lambda)$ is called a Q-neighbourhood base of x_λ, if for every $u \in \mathcal{Q}(x_\lambda)$, there exists $v \in \mathcal{A}$ such that $v \leq u$.

Definition 7.17 [2] Let (L^X, δ) be an L-ts and $x_\lambda \in M(L^X)$. Then the local characteristic $chr_\delta(x_\lambda)$ of x_λ in (L^X, δ) is defined by $chr_\delta(x_\lambda) = \min\{|\mathcal{B}| : \mathcal{B}$ is a Q-neighbourhood base of x_λ in $(L^X, \delta)\}$ and the characteristic $chr(\delta)$ of (L^X, δ) is defined by $chr(\delta) = \vee\{chr_\delta(x_\lambda) : x_\lambda \in M(L^X)\}$.

Definition 7.18 [2] Let (L^X, δ) be an L-ts, (L^X, δ) is called first countable or C_I, if $chr(\delta) \leq \omega$.

Theorem 7.28 [4] *Let (L^X, δ) be an L-ts invertible with respect to an open crisp subset Y. If $(L^Y, \delta|_Y)$ is first countable, then (L^X, δ) is also first countable.*

Proof Let $x_\lambda \in M(L^X)$ and $u \in \mathcal{Q}_\delta(x_\lambda)$. Let $\theta^\rightarrow$ be an inverting homeomorphism for Y so that $Y' \leq \theta^\rightarrow(Y)$. Now there arise two cases:
Case 1: $x_\lambda \in M(L^Y)$. Since $(Y, \delta|_Y)$ is first countable, there exists a Q-neighbourhood base $\mathcal{B}_1$ of x_λ in Y such that $|\mathcal{B}_1| \leq \omega$. We have $u \wedge Y \in \mathcal{Q}_{\delta|_Y}(x_\lambda)$ so that there exists a $v \in \mathcal{B}_1$ such that $v \leq u \wedge Y \leq u$.
Case 2: $x_\lambda \in M(L^{\theta^\rightarrow(Y)})$. Since $(L^{\theta^\rightarrow(Y)}, \delta|_{\theta^\rightarrow(Y)})$ is also first countable, there exists a Q-neighbourhood base $\mathcal{B}_2$ of x_λ in $\theta^\rightarrow(Y)$ such that $|\mathcal{B}_2| \leq \omega$. Since $u \wedge \theta^\rightarrow(Y) \in \mathcal{Q}_{\delta|_{\theta^\rightarrow(Y)}}(x_\lambda)$, there exists a $t \in \mathcal{B}_2$ such that $t \leq u \wedge \theta^\rightarrow(Y) \leq u$.
Hence, either $\mathcal{B}_1$ or $\mathcal{B}_2$ is a Q-neighbourhood base for x_λ in (L^X, δ) so that (L^X, δ) is first countable. $\qquad \square$

Definition 7.19 [2] Let (L^X, δ) be an L-ts and $A \in L^X$. Then the density $dn(\delta)$ of (L^X, δ) is defined by $dn(\delta) = \min\{|A| : A \in L^X, A^- = \underline{1}\}$.

Definition 7.20 [2] Let (L^X, δ) be an L-ts, (L^X, δ) is called separable, if $dn(\delta) \leq \omega$.

Theorem 7.29 [4] *Let (L^X, δ) be an L-ts invertible with respect to an open crisp subset Y. If $(L^Y, \delta|_Y)$ is separable, then (L^X, δ) is also separable.*

Proof Let $\theta^\to$ be an inverting homeomorphism for Y so that $Y' \le \theta^\to(Y)$. Since $(L^Y, \delta|_Y)$ is separable, there exists u in L^Y such that $|u| \le \omega$ and $u^- = 1_Y$. Now consider $\theta^\to(u) \in L^{\theta^\to(Y)}$. We claim $\theta^\to(u)^- = 1_{\theta^\to(Y)}$. For, if $\theta^\to(u)^- = v \ne 1_{\theta^\to(Y)}$, then $\theta^\gets(v) \ne 1_Y$ is a closed L-subset of Y containing u, a contradiction. Hence $t = u \vee \theta^\to(u) \in L^X$ is such that $t^- = 1_X$ and $|t| \le \omega$. □

Definition 7.21 [2] Let (L^X, δ) be an L-ts. (L^X, δ) is called quasi-T_0, if for every two distinguished molecules x_λ and x_γ in (L^X, δ) with same support point x, there exists $u \in \mathcal{Q}_\delta(x_\lambda)$ such that x_γ is not quasi-coincident with u, or there exists $v \in \mathcal{Q}_\delta(x_\gamma)$ such that x_λ is not quasi-coincident with v.

Theorem 7.30 [4] *Let (L^X, δ) be an L-ts invertible with respect to an open crisp subset Y. If $(L^Y, \delta|_Y)$ is quasi-T_0, then (L^X, δ) is also quasi-T_0.*

Proof Let x_λ, x_γ be two distinguished molecules in (L^X, δ) with the same support point x. Let $\theta^\to$ be an inverting homeomorphism of Y. Then by Theorem 7.1, $Y' \le \theta^\to(Y)$.

Case 1: $x_\lambda, x_\gamma \in M(L^Y)$. Since $(L^Y, \delta|_Y)$ is quasi-T_0, there exists $u \in \mathcal{Q}_{\delta|_Y}(x_\lambda)$ such that x_γ is not quasi-coincident with u, or there exists $v \in \mathcal{Q}_{\delta|_Y}(x_\gamma)$ such that x_λ is not quasi-coincident with v. Since Y is open in (L^X, δ), both u and v are in δ. Obviously $u \in \mathcal{Q}_\delta(x_\lambda)$ and $v \in \mathcal{Q}_\delta(x_\gamma)$. Hence (L^X, δ) is also quasi-T_0. *Case 2:* x_λ, $x_\gamma \in M(L^{\theta^\to(Y)})$. Since $(L^{\theta^\to(Y)}, \delta|_{\theta^\to(Y)})$ is also quasi-T_0, this is similar to case 1. □

Definition 7.22 [2] Let (L^X, δ) be an L-ts. (L^X, δ) is called sub-T_0, if for every two distinguished ordinary points x, y in X, there exists $\lambda \in M(L)$ such that there exists $u \in \mathcal{Q}_\delta(x_\lambda)$ and y_λ is not quasi-coincident with u, or there exists $v \in \mathcal{Q}_\delta(y_\lambda)$ and x_λ is not quasi-coincident with v.

Theorem 7.31 [4] *Let (L^X, δ) be an L-ts invertible with respect to an open crisp subset Y. If $(L^Y, \delta|_Y)$ is sub-T_0, then (L^X, δ) is also sub-T_0.*

Proof [4] Let $\theta^\to$ be an inverting homeomorphism for Y so that $Y' \le \theta^\to(Y)$. Let $(L^Y, \delta|_Y)$ be sub-T_0. Then $(\theta^\to(Y), \delta|_{\theta^\to(Y)})$ is also sub-T_0. Let x and y be two distinguished points in X. Then the following three cases arise:

Case 1: $x, y \in Y$. Since $(L^Y, \delta|_Y)$ is sub-T_0, there exist $\lambda \in M(L)$ such that there exists $u \in \mathcal{Q}_{\delta|_Y}(x_\lambda)$ and y_λ is not quasi-coincident with u, or there exists $v \in \mathcal{Q}_{\delta|_Y}(y_\lambda)$ and x_λ is not quasi-coincident with v. Then, since Y is open, both u and v are in δ and $u \in \mathcal{Q}_\delta(x_\lambda)$ and $v \in \mathcal{Q}_\delta(y_\lambda)$. Hence (L^X, δ) is sub-T_0.

Case 2: $x, y \in \theta^\to(Y)$. Since $(L^{\theta^\to(Y)}, \delta|_{\theta^\to(Y)})$ is also sub-T_0, this is similar to case 1.

Case 3: Exactly one of $x, y \in Y$. Without loss of generality, let $x \in Y$ and $y \notin Y$. Then $Y \in \mathcal{Q}_\delta(x_\lambda)$ and y_λ is not quasi-coincident with Y so that (L^X, δ) is sub-T_0. □

Definition 7.23 [2] Let (L^X, δ) be an L-ts. (L^X, δ) is called T_0, if for every two distinguished molecules x_λ and y_γ in (L^X, δ), there exists $u \in Q_\delta(x_\lambda)$ such that y_γ is not quasi-coincident with u, or there exists $v \in Q_\delta(y_\gamma)$ such that x_λ is not quasi-coincident with v.

Theorem 7.32 [4] *Let (L^X, δ) be an L-ts invertible with respect to an open crisp subset Y. If $(L^Y, \delta|_Y)$ is T_0, then (L^X, δ) is also T_0.*

Proof Let $\theta^\rightarrow$ be an inverting homeomorphism for Y so that $Y' \leq \theta^\rightarrow(Y)$. Let $(L^Y, \delta|_Y)$ be T_0. Then $(\theta^\rightarrow(Y), \delta|_{\theta^\rightarrow(Y)})$ is also T_0. Let x_λ and y_γ be two distinguished molecules in (L^X, δ). Then the following three cases arise:

Case 1: $x_\lambda, y_\gamma \in M(L^Y)$. Since $(L^Y, \delta|_Y)$ is T_0, there exists $u \in Q_{\delta|_Y}(x_\lambda)$ and y_γ does not quasi-coincident with u, or there exists $v \in Q_{\delta|_Y}(y_\gamma)$ and x_λ is not quasi-coincident with v. Since Y is open, it follows that $u \in Q_\delta(x_\lambda)$ and $v \in Q_\delta(y_\gamma)$. Hence (L^X, δ) is T_0. *Case 2*: $x_\lambda, y_\gamma \in M(L^{\theta^\rightarrow(Y)})$. Since $(L^{\theta^\rightarrow(Y)}, \delta|_{\theta^\rightarrow(Y)})$ is also T_0, this is similar to case 1. *Case 3*: Exactly one of $x_\lambda, y_\gamma \in M(L^Y)$. Without loss of generality, let $x_\lambda \in M(L^Y)$ and $y_\gamma \notin M(L^Y)$. Then $Y \in Q_\delta(x_\lambda)$ and y_γ is not quasi-coincident with Y. Consequently (L^X, δ) is T_0. $\square$

Definition 7.24 [2] Let (L^X, δ) be an L-ts. (L^X, δ) is called T_1, if for every two distinguished molecules m and n in (L^X, δ) such that $m \not\leq n$, there exists $u \in Q_\delta(m)$ such that n is not quasi-coincident with u.

Theorem 7.33 [4] *Let (L^X, δ) be an L-ts invertible with respect to an open crisp subset Y. If $(L^Y, \delta|_Y)$ is T_1, then (L^X, δ) is also T_1.*

Proof Let $\theta^\rightarrow$ be an inverting homeomorphism for Y so that $Y' \leq \theta^\rightarrow(Y)$. Let $(L^Y, \delta|_Y)$ be T_1. Then $(\theta^\rightarrow(Y), \delta|_{\theta^\rightarrow(Y)})$ is also T_1. Let m and n be two distinguished molecules in (L^X, δ) such that $m \not\leq n$. Then the following three cases arise:
Case 1: $m, n \in M(L^Y)$. Since $(L^Y, \delta|_Y)$ is T_1, there exists $u \in Q_{\delta|_Y}(m)$, such that n is not quasi-coincident with u. Since Y is open, it follows that $u \in Q_\delta(m)$. Hence (L^X, δ) is T_1.
Case 2: $m, n \in M(L^{\theta^\rightarrow(Y)})$. Since $(L^{\theta^\rightarrow(Y)}, \delta|_{\theta^\rightarrow(Y)})$ is also T_1, this is similar to case 1.
Case 3: Exactly one of $m, n \in M(L^Y)$. Without loss of generality, let $m \in M(L^Y)$. Then $Y \in Q_\delta(m)$ and n is not quasi-coincident with Y. Consequently (L^X, δ) is T_1. $\square$

Theorem 7.34 [2] *Let (L^X, δ) be an L-ts. Then (L^X, δ) is typically T_1 if and only if for every $x_\lambda \in Pt(L^X)$ and every $m \in M(L^X)$ such that $m \not\leq x_\lambda$, there exists $u \in Q(m)$ such that x_λ is not quasi-coincident with u.*

Theorem 7.35 [4] *Let (L^X, δ) be an L-ts invertible with respect to an open crisp subset Y. If $(L^Y, \delta|_Y)$ is typically T_1, then (L^X, δ) is also typically T_1.*

Proof Let $\theta^{\to}$ be an inverting homeomorphism for Y so that $Y' \leq \theta^{\to}(Y)$. Let $(L^Y, \delta|_Y)$ be typically T_1. Then $(\theta^{\to}(Y), \delta|_{\theta^{\to}(Y)})$ is also typically T_1. Let x_λ be an L-point in X and m be a molecule in X such that $m \not\leq x_\lambda$. Then the following four cases arise:

Case 1: $x_\lambda \in Pt(L^Y)$ and $m \in M(L^Y)$. Since $(L^Y, \delta|_Y)$ is typically T_1 by Theorem 7.34, there exists $u \in \mathcal{Q}_{\delta|_Y}(m)$ such that x_λ is not quasi-coincident with u. Since Y is open, it follows that $u \in \mathcal{Q}_\delta(m)$.

Case 2: $x_\lambda \in Pt(L^{\theta^{\to}(Y)})$ and $m \in M(L^{\theta^{\to}(Y)})$. Since $(L^{\theta^{\to}(Y)}, \delta|_{\theta^{\to}(Y)})$ is also typically T_1, this is similar to case 1.

Case 3: $x_\lambda \in Pt(L^Y)$, $x_\lambda \notin Pt(L^{\theta^{\to}(Y)})$ and $m \in M(L^{\theta^{\to}(Y)})$, $m \notin M(L^Y)$. Then $\theta^{\to}(Y) \in \mathcal{Q}_\delta(m)$ and x_λ is not quasi-coincident with $\theta^{\to}(Y)$.

Case 4: $x_\lambda \in Pt(L^{\theta^{\to}(Y)})$, $x_\lambda \notin Pt(L^Y)$ and $m \in M(L^Y)$, $m \notin M(L^{\theta^{\to}(Y)})$. Then $Y \in \mathcal{Q}_\delta(m)$ and x_λ is not quasi-coincident with Y.

Thus, in all the above cases, there exists a Q-neighbourhood u of m in δ such that x_λ is not quasi-coincident with u so that by Theorem 7.34, (L^X, δ) is typically T_1. $\square$

Definition 7.25 [2] Let (L^X, δ) be an L-ts. (L^X, δ) is called T_2, if for every two molecules x_λ and y_γ in (L^X, δ) with distinguished support points $x \neq y$, there exist $u \in \mathcal{Q}_\delta(x_\lambda)$ and $v \in \mathcal{Q}_\delta(y_\gamma)$ such that $u \wedge v = \underline{0}$.

Theorem 7.36 [4] *Let (L^X, δ) be a completely invertible L-ts and Y be a proper open crisp subset of X. If $(L^Y, \delta|_Y)$ is T_2, then (L^X, δ) is also T_2.*

Proof Let $\theta^{\to}$ be an inverting homeomorphism for Y so that $Y' \leq \theta^{\to}(Y)$. Let $(L^Y, \delta|_Y)$ be T_2. Then $(\theta^{\to}(Y), \delta|_{\theta^{\to}(Y)})$ is also T_2. Let x_λ and y_γ be two molecules in (L^X, δ) with support points $x \neq y$. Then the following three cases arise:

Case 1: $x, y \in Y$. Since $(L^Y, \delta|_Y)$ is T_2, there exists $u \in \mathcal{Q}_{\delta|_Y}(x_\lambda)$ and $v \in \mathcal{Q}_{\delta|_Y}(y_\gamma)$ such that $u \wedge v = \underline{0}$. Since Y is open in (L^X, δ), it follows that $u \in \mathcal{Q}_\delta(x_\lambda)$ and $v \in \mathcal{Q}_\delta(y_\gamma)$. Case 2: $x, y \in \theta^{\to}(Y)$. Since $(L^{\theta^{\to}(Y)}, \delta|_{\theta^{\to}(Y)})$ is also T_2, this is similar to case 1.

Case 3: $x \in Y$, $x \notin \theta^{\to}(Y)$ and $y \in \theta^{\to}(Y)$, $y \notin Y$.

If $Y \wedge \theta^{\to}(Y) = \underline{0}$, then since $Y \in \mathcal{Q}_\delta(x_\lambda)$ and $\theta^{\to}(Y) \in \mathcal{Q}_\delta(y_\gamma)$, the result follows. Otherwise, let $u = Y \wedge \theta^{\to}(Y)$. Clearly $x, y \in u'$. Let $\eta^{\to}$ be an inverting homeomorphism for u. Then $\eta^{\to}(x_\lambda)$, $\eta^{\to}(y_\gamma) \in u \leq Y$ and $\eta(x) \neq \eta(y)$. Now by case 1, there exists $p \in \mathcal{Q}_\delta(\eta^{\to}(x_\lambda))$ and $q \in \mathcal{Q}_\delta(\eta^{\to}(y_\gamma))$ such that $p \wedge q = \underline{0}$. Thus there exist $\eta^{\leftarrow}(p) \in \mathcal{Q}_\delta(x_\lambda)$ and $\eta^{\leftarrow}(q) \in \mathcal{Q}_\delta(y_\gamma)$ such that $\eta^{\leftarrow}(p) \wedge \eta^{\leftarrow}(q) = \underline{0}$. Consequently (L^X, δ) is T_2. $\square$

Definition 7.26 [2] Let (L^X, δ) be an L-ts, $A \in L^X, u \in \delta$. u is called quasi-coincident neighbourhood of A, or a Q-neighbourhood of A for short, if A quasi-coincides with u at every $x \in supp\, A$. Denote the family of all the Q-neighbourhoods of A in (L^X, δ) by $\mathcal{Q}_\delta(A)$, or $\mathcal{Q}(A)$ for short.

Definition 7.27 [2] (L^X, δ) is called pararegular or p-regular for short, if for every non-zero pseudo-crisp closed subset k in (L^X, δ) and every molecule $x_\lambda \in M(L^X)$ such that $x \notin supp\, k$, there exist $u \in \mathcal{Q}(x_\lambda)$ and $v \in \mathcal{Q}(k)$ such that $u \wedge v = \underline{0}$.

Theorem 7.37 [4] *Let (L^X, δ) be a completely invertible L-ts. If $(Y, \delta|_Y)$ is p-regular, then (L^X, δ) is also p-regular.*

Proof Let $x_\lambda \in M(L^X)$ and k be a pseudo-crisp closed L-subset in (L^X, δ) such that $x \notin supp\, k$. Now consider the following three cases:

Case 1: $x_\lambda \notin M(L^Y)$ and $Y \wedge k' \neq \underline{0}$. Let $\theta^\rightarrow$ be an inverting homeomorphism for $Y \wedge k'$. Then $\theta^\rightarrow(x_\lambda) \in M(L^Y)$ and $\theta^\rightarrow(k) \leq Y \wedge k' \leq Y$ where $\theta^\rightarrow(k)$ is closed in Y. Since $(Y, \delta|_Y)$ is p-regular and $\theta^\rightarrow(x) \notin supp\, \theta^\rightarrow(k)$, there exist $u \in \mathcal{Q}_{\delta|_Y}(\theta^\rightarrow(x_\lambda))$ and $v \in \mathcal{Q}_{\delta|_Y}(\theta^\rightarrow(k))$ such that $u \wedge v = \underline{0}$. Since Y is open, $u \in \mathcal{Q}_\delta(\theta^\rightarrow(x_\lambda))$ and $v \in \mathcal{Q}_\delta(\theta^\rightarrow(k))$ so that $\theta^\leftarrow(u) \in \mathcal{Q}_\delta(x_\lambda)$ and $\theta^\leftarrow(v) \in \mathcal{Q}_\delta(k)$. Also $\theta^\leftarrow(u) \wedge \theta^\leftarrow(v) = \underline{0}$.

Case 2: $x_\lambda \notin M(L^Y)$ and $Y \wedge k' = \underline{0}$. Let $\theta^\rightarrow$ be an inverting homeomorphism for Y. Then $\theta^\rightarrow(x_\lambda) \in Y$ and $\theta^\rightarrow(k \wedge Y') \leq Y$. Also $\theta^\rightarrow(k \wedge Y')$ is closed in Y and $\theta^\rightarrow(x) \notin supp\, \theta^\rightarrow(k \wedge Y')$. Since $(Y, \delta|_Y)$ is p-regular, there exist $u \in \mathcal{Q}_{\delta|_Y}(\theta^\rightarrow(x_\lambda))$ and $v \in \mathcal{Q}_{\delta|_Y}(\theta^\rightarrow(k \wedge Y'))$ such that $u \wedge v = \underline{0}$. Since Y is open, $u \in \mathcal{Q}_\delta(\theta^\rightarrow(x_\lambda))$ and $v \in \mathcal{Q}_\delta(\theta^\rightarrow(k \wedge Y'))$. Consequently, $\theta^\leftarrow(u) \in \mathcal{Q}_\delta(x_\lambda)$ and $\theta^\leftarrow(v) \in \mathcal{Q}_\delta(k \wedge Y')$ such that $\theta^\leftarrow(u) \wedge \theta^\leftarrow(v) = \underline{0}$. But $k \leq (k \wedge Y') \vee Y$ so that $\theta^\leftarrow(v) \vee Y \in \mathcal{Q}_\delta(k)$. Also we have $\theta^\leftarrow(u) \wedge k' \in \mathcal{Q}_\delta(x_\lambda)$. Obviously, $(\theta^\leftarrow(u) \wedge k') \wedge (\theta^\leftarrow(v) \vee Y) = \underline{0}$.

Case 3: $x_\lambda \in M(L^Y)$. Let $\theta^\rightarrow$ be an inverting homeomorphism for Y.

Subcase 3.1: Suppose $\theta^\rightarrow(x_\lambda) \in Y'$. We have $\theta^\rightarrow(k)$ is closed in X and $\theta^\rightarrow(x) \notin supp\, \theta^\rightarrow(k)$. Since $\theta^\rightarrow(x_\lambda) \notin M(L^Y)$, by case 1 and case 2, there exist $u \in \mathcal{Q}_\delta(\theta^\rightarrow(x_\lambda))$ and $v \in \mathcal{Q}_\delta(\theta^\rightarrow(k))$ with $u \wedge v = \underline{0}$. Then there exist $\theta^\leftarrow(u) \in \mathcal{Q}_\delta(x_\lambda)$ and $\theta^\leftarrow(v) \in \mathcal{Q}_\delta(k)$ such that $\theta^\leftarrow(u) \wedge \theta^\leftarrow(v) = \underline{0}$.

Subcase 3.2: Suppose $\theta^\rightarrow(x_\lambda) \in Y$. Since $k \wedge Y$ is closed in Y and $x \notin supp\, (k \wedge Y)$, there exist $u \in \mathcal{Q}_{\delta|_Y}(x_\lambda)$ and $v \in \mathcal{Q}_{\delta|_Y}(k \wedge Y)$ with $u \wedge v = \underline{0}$. But $\theta^\rightarrow(k \wedge Y') \leq Y$ and $\theta^\rightarrow(x_\lambda) \in Y$. Also we have $\theta^\rightarrow(x) \notin supp\, (\theta^\rightarrow(k \wedge Y'))$ so that there exist $p \in \mathcal{Q}_{\delta|_Y}(\theta^\rightarrow(x_\lambda))$ and $q \in \mathcal{Q}_{\delta|_Y}(\theta^\rightarrow(k \wedge Y'))$ with $p \wedge q = \underline{0}$. Then $\theta^\leftarrow(p) \in \mathcal{Q}_\delta(x_\lambda)$ and $\theta^\leftarrow(q) \in \mathcal{Q}_\delta(k \wedge Y')$ such that $\theta^\leftarrow(p) \wedge \theta^\leftarrow(q) = \underline{0}$. Thus $u \wedge \theta^\leftarrow(p) \in \mathcal{Q}_\delta(x_\lambda)$ and $v \vee \theta^\leftarrow(q) \in \mathcal{Q}_\delta((k \wedge Y) \vee (k \wedge Y')) = \mathcal{Q}_\delta(k)$ such that $(u \wedge \theta^\leftarrow(p)) \wedge (v \vee \theta^\leftarrow(q)) = \underline{0}$. Consequently (L^X, δ) is p-regular. $\qquad\square$

Definition 7.28 [2] Let (L^X, δ) be an L-ts. (L^X, δ) is called normal, if for every closed L-subset w and every open L-subset u in (L^X, δ) such that $w \leq u$, there exists an open L-subset v in (L^X, δ) such that $w \leq v \leq v^- \leq u$.

Theorem 7.38 [2] *Let (L^X, δ) be an L-ts. Then (L^X, δ) is normal if and only if for every two closed L-subsets p and q in (L^X, δ) such that p is not quasi-coincident with q, there exists open L-subsets u and v in $L^X, \delta)$ such that $p \leq u$ and $q \leq v$ and u is not quasi-coincident with v.*

Theorem 7.39 [5] *Let (L^X, δ) be an L-ts invertible with respect to an open crisp subset Y. If $(L^Y, \delta|_Y)$ is normal, then (L^X, δ) is also normal.*

Proof Let p and q be two closed L-subsets of X such that p is not quasi-coincident with q, i.e. $p \leq q'$. Let $\theta^\rightarrow$ be an inverting homeomorphism for Y. Since $(L^Y, \delta|_Y)$ is normal $(L^{\theta^\rightarrow(Y)}, \delta|_{\theta^\rightarrow(Y)})$ is also normal. Let $Y_1 = Y \wedge \theta^\rightarrow(Y)$. Then we have $p = (p \wedge Y_1) \vee (p \wedge Y_1') = [p \wedge \theta^\rightarrow(Y)'] \vee [p \wedge (Y \wedge \theta^\rightarrow(Y))] \vee$

$[p \wedge Y'] = m_1 \vee m_2 \vee m_3$ where $m_1 = p \wedge (\theta^\rightarrow(Y))'$, $m_2 = p \wedge (Y \wedge \theta^\rightarrow(Y))$ and $m_3 = p \wedge Y'$. Similarly, $q = [q \wedge (\theta^\rightarrow(Y))'] \vee [q \wedge (Y \wedge \theta^\rightarrow(Y))] \vee [q \wedge Y'] = n_1 \vee n_2 \vee n_3$ where $n_1 = q \wedge (\theta^\rightarrow(Y))'$, $n_2 = q \wedge (Y \wedge \theta^\rightarrow(Y))$ and $n_3 = q \wedge Y'$. We have $m_1 \vee m_2$ and $n_1 \vee n_2$ are closed *L*-subsets of Y. For, $m_1 \vee m_2 = p \wedge Y$ and $n_1 \vee n_2 = q \wedge Y$. Since $(Y, \delta|_Y)$ is normal and $m_1 \vee m_2 \leq (n_1 \vee n_2)'$, by Theorem 7.38, there exist u_1, $u_2 \in \delta|_Y$ such that $m_1 \vee m_2 \leq u_1$ and $n_1 \vee n_2 \leq u_2$ with $u_1 \leq u_2'$. Similarly, $m_2 \vee m_3 = p \wedge \theta^\rightarrow(Y)$ and $n_2 \vee n_3 = q \wedge \theta^\rightarrow(Y)$ are closed *L*-subsets of $\theta^\rightarrow(Y)$ with $m_2 \vee m_3 \leq (n_2 \vee n_3)'$. Since $(L^{\theta^\rightarrow(Y)}, \delta|_{\theta^\rightarrow(Y)})$ is normal, there exist v_1, $v_2 \in \delta_{\theta^\rightarrow(Y)} \subset \delta$ such that $m_2 \vee m_3 \leq v_1$ and $n_2 \vee n_3 \leq v_2$ with $v_1 \leq v_2'$.

Now we claim that there are open *L*-subsets u_3, u_4, v_3 and v_4 in X such that $m_3 \leq u_3$ and $n_1 \leq u_4$ with $u_3 \leq (u_4)'$ and $m_1 \leq v_3$ and $n_3 \leq v_4$ with $v_3 \leq v_4'$. For, if $Y_1 = \underline{0}$ then $u_4 = v_3 = Y$ and $u_3 = v_4 = \theta^\rightarrow(Y)$ will suffice. Otherwise, let $\eta^\rightarrow$ be an inverting homeomorphism for Y_1. Clearly, $\eta^\rightarrow(m_1)$, $\eta^\rightarrow(m_3)$, $\eta^\rightarrow(n_1)$, $\eta^\rightarrow(n_3) \leq Y_1 \leq Y$ and are closed in Y such that $\eta^\rightarrow(m_3) \leq (\eta(n_1))'$ and $\eta^\rightarrow(m_1) \leq (\eta^\rightarrow(n_3))'$. Then since $(Y, \delta|_Y)$ is normal, there exist k, l, r, $s \in \delta_Y \subset \delta$ such that $\eta^\rightarrow(m_3) \leq k$ and $\eta^\rightarrow(n_1) \leq l$ with $k \leq l'$ and $\eta^\rightarrow(m_1) \leq r$ and $\eta^\rightarrow(n_3) \leq s$ with $r \leq s'$. Now taking $u_3 = \eta^{-1}(k)$, $u_4 = \eta^{-1}(l)$, $v_3 = \eta^{-1}(r)$ and $v_4 = \eta^{-1}(s)$, we get open *L*-subsets as claimed.

Clearly $p \leq u_1 \vee u_3$ and $n_1 \leq u_2 \wedge u_4$ with $u_1 \vee u_3 \leq (u_2 \wedge u_4)'$. Also $p \leq u_1 \vee v_1$ and $n_2 \leq u_2 \wedge v_2$ with $u_1 \vee v_1 \leq (u_2 \wedge v_2)'$. Further, $p \leq v_1 \vee v_3$ and $n_3 \leq v_2 \wedge v_4$ with $v_1 \vee v_3 \leq (v_2 \wedge v_4)'$. Finally, let $u = (u_1 \vee u_3) \wedge (u_1 \vee v_1) \wedge (v_1 \vee v_3)$ and $v = (u_2 \wedge u_4) \vee (u_2 \wedge v_2) \vee (v_2 \wedge v_4)$. Then u, $v \in \delta$ such that $p \leq u$ and $q \leq v$ with $u \leq v'$. Hence by Theorem 7.38, (L^X, δ) is normal. $\square$

Remark 7.13 It is easy to illustrate that the invertibility or complete invertibility is essential for the above results.

7.5   Exercises

7.1 Prove or disprove: an *L*-ts with a closed weak *L*-point is invertible.

7.2 Show that every H-*L*-ts containing a well-closed *L*-point is invertible.

7.3 Show that every H-*L*-ts with a closed *L*-point is invertible.

7.4 Let (L^X, δ) be a completely invertible *L*-ts and $X_0 = X$. Then prove that (L^X, δ) is an H-*L*-ts.

7.5 Give an example for an invertible L-ts where the six inverting pairs mentioned in Theorem 7.5 are different.

7.6 Show that the sum space of a family of completely invertible *L*-ts need not be completely invertible.

7.7 Prove that the support of an inverting L-subset, if open is also an inverting set.

7.8 Prove that a finite completely invertible c-L-ts is not connected.

7.9 Show by examples that complete invertibility cannot be dropped from (a) Theorem 7.36, (b) Theorem 7.37 and (c) Theorem 7.39.

7.10 Prove that the product topology (X, δ) of a family of L-topologies (X_t, δ_t), $t \in T$ is invertible if (X_t, δ_t) is invertible for some $t \in T$.

7.11 Show that the converse of Exercise 7.10 is not true.

7.12 Prove that the product L-topology (X, δ) of a family of L-ts' (X_t, δ_t), $t \in \delta$ is type 1 invertible if (X_t, δ_t) is type 1 invertible for some $t \in T$.

7.13 Show by an example that the converse of Exercise 7.12 is not true.

7.14 Prove that the product of a family of type 1 invertible L-topological spaces is type 1 invertible.

7.15 If the product topology (X, δ) of a family of invertible L-ts' (X_t, δ_t), $t \in T$ is type 2 invertible, then show that each (X_t, δ_t) is type 2 invertible.

7.16 Show that the product L-topology (X, δ) of a family of L-ts' (X_t, δ_t), $t \in T$ is type 2 completely invertible if and only if (X_t, δ_t) is type 2 completely invertible for each $t \in T$.

7.17 Prove that the product of a family of invertible L-topological spaces is invertible.

7.18 Prove by example that the product space may be completely invertible even if none of the coordinate spaces is invertible.

References

1. Jose, A., Mathew, S.C.: Invertibility in L-topological spaces. Fuzzy Inf. Eng. **6**, 41–57 (2014)
2. Liu, Y.M., Luo, M.K.: Fuzzy Topology. World Scientific (1997)
3. Mathew, S.C.: Adjacent L-fuzzy topological spaces. Int. J. Appl. Math. Stat. **7**, 1–13 (2006)
4. Jose, A., Mathew, S.C.: Local to global properties of invertible l topologies. Anals. U. Oradea-Fasicola Mathematica **2**, 85–94 (2013)
5. Jose, A.: Some properties of invertible fuzzy topological spaces and related topics. Ph.D. thesis, Mahatma Gandhi University, Kottayam, Kerala, India (2013)

Index

© The Editor(s) (if applicable) and The Author(s), under exclusive license
to Springer Nature Singapore Pte Ltd. 2022
A. Jose and S. C. Mathew, *Invertible Fuzzy Topological Spaces*,
https://doi.org/10.1007/978-981-19-3689-0

Printed in the United States
by Baker & Taylor Publisher Services

Printed in the United States
by Baker & Taylor Publisher Services